DE INTESTINI DUODENI SITU ET NEXU

of Laurentius Claussen

A Dissertation on the Topographical
Anatomy of the Duodenum from 1757

Translated & Annotated by
CLAUDIO GUERRIERI

CONTENTS

INTRODUCTION

It is the middle of the 18th century and a medical student at the University of Leipzig in Germany must come up with a topic of investigation for his thesis. After much deliberation, he eventually decides to study and discuss the topographical anatomy of the duodenum.

But why the duodenum?

This organ, or better yet, this first and shortest segment of the small intestine had been misunderstood for centuries. Its documented story began at the time of Herophilus in the 3rd century BC, and its long story is rife with confusion. Certain authors were struck by selective amnesia and denied its existence. Others could not make sense of its name — a name which referred to its length of twelve "finger breadths" or, erroneously, twelve inches. Some have said that the duodenum begins at the pylorus and ends at the insertion of the common bile duct and, therefore, corresponded to the much shorter length of 4-6 inches. Based on this, there was a period in which some anatomists extrapolated the absurd notion that 2,000 years ago humans must have been much taller and believed that our human ancestors were 10-foot-tall giants! There were those who depicted the duodenum as a straight tube. Others considered it as part of the stomach, or even an accessory stomach after being misled by the excessive dilatation that it would sometimes attain.

Through this haze of confusing interpretations, the anatomists of the 18th century had to straighten things out with their own personal and critical observations. It was in this context that Claussen's thesis would come along and accurately describe the anatomy of this organ and, most importantly, present two new and truthful drawings of the duodenum as it lies within the human body.

LAURENTIUS CLAUSSEN

Little is known of the personal life of Lorenz Claussen (1725-1771), but his academic career was surprisingly well documented. His father was the Danish surgeon Henning Ditlev Claussen (ca. 1690-1734). Henning belonged to a group of free surgeons who practiced without any formal training or qualifications in the Copenhagen of the early 18th century. Most of these surgeons were of dubious reputation, but Henning was held in high esteem for his skills as a general surgeon and in midwifery. The Guild had tried to oppose his career but seemingly without success. Henning was a student of Johannes de Buchwald (1658-1738), a noted surgeon and professor of medicine, anatomy, botany and surgery at the University of Copenhagen, and the first to give lectures of obstetrics in Denmark. Henning became one of the most sought-after surgeons of Copenhagen and at one time was the only one to have a major obstetrical practice in the city.

Henning Claussen married Agnete Schiøtt (1709-1732) who came from a reputable family. Her parents were the grocer Balthasar Paulsen Schiøtt and his wife Elisa (or Else) Maria.

Henning and Agnete had two sons. Their eldest son, Lorenz, was born in Copenhagen in 1725. He pursued a legitimate medical education which he began in 1748. Lorenz first studied in Copenhagen and then abroad. In 1753, he was in Berlin and thereafter transferred to Leipzig in 1755 where he obtained his medical degree in the year 1757.

The name for Lorenz Claussen changes somewhat among the various bibliographical references. For instance, the surname Claussen is written sometimes as "Clausen." The first name "Laurentius" is the latinized version used for publications, and his first name is found as Lorenz, Lorentz or Laurinds. His father's name also varies, with his middle name Ditlev sometimes written as "Ditlef," and his family name as "Clausen." For the sake of consistency, the "Lorenz" version of his name will be used from here on within this book.

In 1756, Lorenz **Claussen** wrote the booklet titled *De Medicamentis ut Menstruum Agentibus ad Leges Chymicas*. This was based on a dissertation he gave on December 11, 1756, held between the hours 9:00-12:00 in the Auditorium of Medicine in the University of Leipzig.

This dissertation was reviewed in the *Neue Zeitungen von Gelehrten Sachen auf das Jahr 1757*:

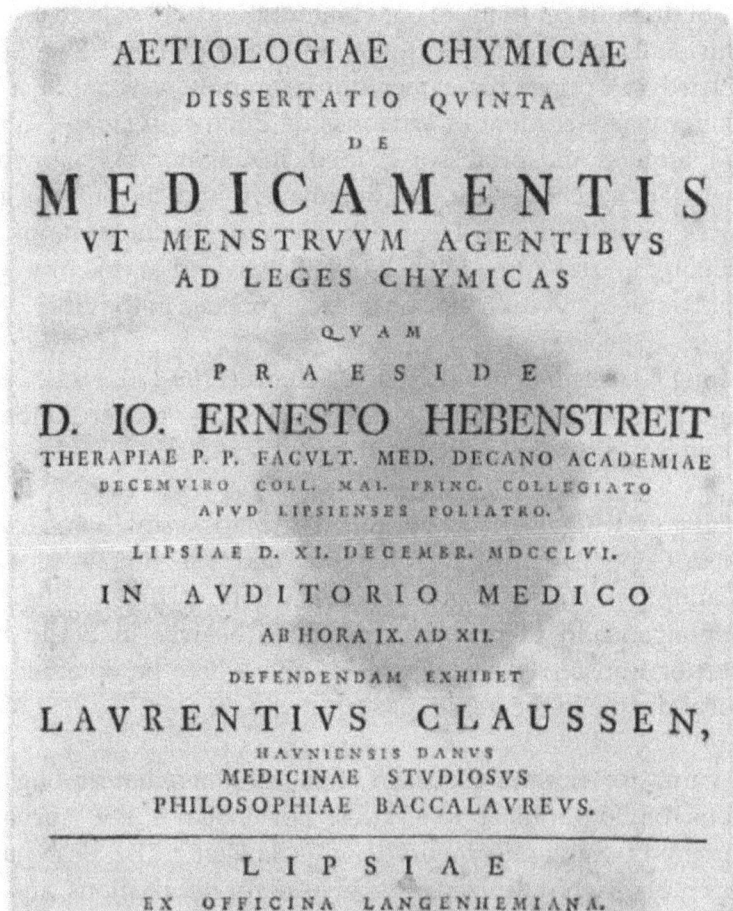

AETIOLOGIAE CHYMICAE

DISSERTATIO QVINTA

D E

MEDICAMENTIS

VT MENSTRVVM AGENTIBVS
AD LEGES CHYMICAS

Q V A M

P R A E S I D E

D. IO. ERNESTO HEBENSTREIT

THERAPIAE P. P. FACVLT. MED. DECANO ACADEMIAE
DECEMVIRO COLL. MAI. PRINC. COLLEGIATO
APVD LIPSIENSES POLIATRO.

LIPSIAE D. XI. DECEMBR. MDCCLVI.

IN AVDITORIO MEDICO

AB HORA IX. AD XII.

DEFENDENDAM EXHIBET

LAVRENTIVS CLAUSSEN,

HAVNIENSIS DANVS
MEDICINAE STVDIOSVS
PHILOSOPHIAE BACCALAVREVS.

L I P S I A E

EX OFFICINA LANGENHEMIANA.

"Im 11ten December, vertheydigte unter dem Vorsitz Herrn D. Johann Ernst Hebenstreits, Herr Laurentius Claussen, von Kopenhagen, Aetiologiae chymicae Dissertationem quintam, de medicamentis, ut menstruum agentibus, ad leges chymicas. Er beschreibt anfanglich was ein menstruum chymicum sey, nehmlich dasjenige Mittel, wodurch ein festes sowohl als ein flütziges Wesen in seine ersten Bestandtheile aufgelöset werden kann, und zeiget hernach, dass es dergleichen menstrua in unserm Körper gebe, nehmlich die Wärme, und deren verschiedene Arten, welche sowohl innerlich in dem Körs per sind, als auch ausserlich an denselben gebracht werden; Ferden ziehet er hieher eine elektrische Kraft, allerhand Arzneymittel, und giebt die Art und Weise an, wie diese verschiedenen menstrua in dem menschlichen Körper wirken. Nicht weniger rechnet er die Luft, verschiedene Wasser, spirituose Sachen, und ühlichte Körper hieher, deren bestimmte Art aufzulösen er sowohl anzeigt, als auch den Nutzen in denen Krankheiten fleissig beschreibet."

[On the 11th of December, under the chairmanship of Mr. D. Johann Ernst Hebenstreit, Mr. Laurentius Claussen, of Copenhagen, defended *Aetiologiae chymicae, Dissertationem quintam, de medicamentis, ut menstruum agentibus, ad leges chymicas*. In the beginning he describes what *menstruum chymicum* {chemical digestion} is, namely a means whereby a solid, as well as a fluid, can be resolved into its primary components, and shows that there is such digestion in our bodies, namely, heat, and their different kinds, which are both internal in the body, and also the manner in which these different digestions occur in the human body. Not only does he include gas, various fluids, and spiritual things, and of course the Body, whose particular kind of digestion he both explains, but also the benefits in those diseases which he diligently describes.]

On October 21, 1757, Claussen defended his dissertation for his medical degree in the University of Leipzig. It was titled *De Intestini Duodeni Situ et Nexu* [On the Position and Connections of the Duodenal Intestine], which basically can be translated as "The Topographical Anatomy of the Duodenum." It was then published in Leipzig as a quarto booklet printed with two copper lithographs.

The book was republished in the 1778 edition of a collection of dissertations called *Thesaurus Dissertationum Programmatum, volume III* and edited by the anatomist Eduard **Sandifort**. Sandifort's presentation to Claussen's dissertation is as follows:

> "Quum nullum inveniret auctorem, qui intestinum duodenum satis accurate expressit, novam delineationem elaborare tentavit, ducto in consortium Clar. Reichelio. Indicat dissectores, qui in duodenum inquisiverunt, difficilem esse hanc disquisitionem & delineationem ostendit, rationem disquirendi proponit, situm & nexum describit, partes, ipsi incumbentes, recenset, limites definit, situm vesiculae felleae juxta duodenum attendit, aliaque ad situm & nexum pertinentia, enarrat, causas digestionis in duodeno concurrentes brevibus indicat, usum duodeni generalem explicat; a partibus vicinis morbo adfectis illud laedi porro ostendit, & nimis extensum in partes vicinas agere debere, probat."

[When no author was found that had depicted the intestine called the *duodenum* with sufficient accuracy, he {Claussen} attempted to elaborate a new drawing, guided in the partnership with the illustrious Reichel. He informs those dissectors who wish to examine the duodenum that it is difficult to perform an examination and drawing of this. Thus, he proposes a method of investigation, describes its position and connections, enumerates the organs that lie adjacent to it, defines its boundaries, pays attention to the location of the gallbladder near the duodenum, reports on other matters pertinent to its position and relation, briefly indicates the simultaneously occurring causes of digestion in the duodenum, and explains the general function of the duodenum. He then shows how this can be damaged by nearby organs affected by disease and proves that when excessively enlarged it will impinge upon other organs.]

The two figures from his original work were reproduced in the 1778 edition and were placed side by side as a folding leaf.

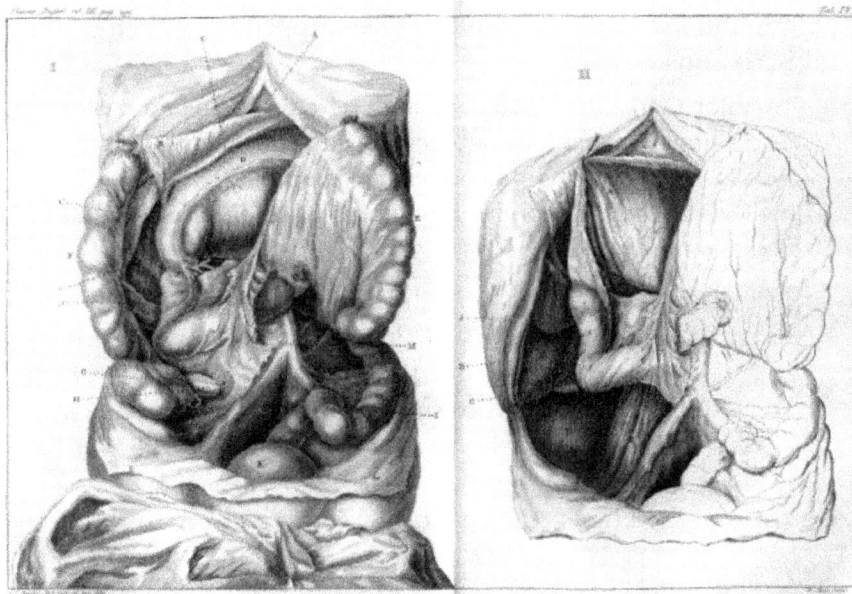

Claussen's drawings (1757) reproduced in Sandifort's volume (1780).

After graduating, Claussen presumably went to Edinburgh where he continued his studies. He then worked as a doctor in the French Army during the Seven Years' War. Towards the end of the war, in 1762, he returned to Denmark and practiced as a physician in Copenhagen. In that same year, he was appointed Court-Physician and remained in this position until his death in 1771. He was married to his cousin Antonette Schiøtt (1740-1779).

Claussen's two illustrations of the duodenum were included in a booklet written by his mentor Professor Johann Ernst **Hebenstreit** of Leipzig. Hebenstreit was a German physician, anatomist, naturalist and traveller, and was born in Neustadt in 1703. He

6

became professor of Surgery (1737), of Pathology (1746) and of Medicine (1748) in the University of Leipzig. In 1757 he wrote a translation of part of the 9th book of the Byzantine-Greek physician Aetius of Amida (Mesopotamia) who lived in the 6th century AD. It was published as *Aetii Amideni Anekdoton liber IX, caput XXVII, exhibens tenuioris intestini morbum quem Ileon et Chordapsum dicunt.* The topic was the *passio iliaca* or ileus of the small bowel. Hebenstreit also included at the end of this dissertation several pages dedicated to the curriculum vitae of Claussen, wherein he praised his student for his in-depth description of the duodenum (see Appendix for the full transcription and notations).

> "Gratulor Clarissimo Candidato meo de erudita huius intestini descriptione, qua effecit, ut omnes intelligant id, quod nos privatis experimentis de eodem captis satis superque intelleximus, eum honoribus academicis proxime secuturis esse dignissimum."

> [I congratulate my very bright candidate on his erudite description of this intestine which he has accomplished so that all may comprehend it, and we may understand more from similar personal observations, and for this he is worthy of the academic honors that will follow him.]

Hebenstreit died that same year of 1757.

One of the most interesting features of Claussen's book is how he collated most of the current knowledge of the duodenum as written by many of the masters of Anatomy of the time.

In the following chapters, the entire manuscript of Claussen's *De Intestini Duodeni Situ et Nexu* is transcribed and each chapter is translated from Latin to English. The original Latin text is placed on the *verso* page, and its corresponding English translation on the *recto* page. For those chapters with footnotes and references embedded at the bottom of the original text, these are examined and amplified in order to document the sources that were drawn upon for Claussen's dissertation. The parentheses {...} indicate a word or words added to complete the meaning of a particular phrase.

AETII AMIDENI

Α Ν Ε Κ Δ Ο Τ Ω Ν

LIB. IX. CAPVT XXVIII.

EXHIBENS

TENVIORIS INTESTINI MORBVM

QVEM

ILEON ET CHORDAPSVM

DICVNT.

VNA CVM

VETERVM SVPER HAC AEGROTATIONE

SENTENTIIS SISTIT

ET

PANEGYRIN MEDICAM

LIPSIAE DIE XXI. OCTOBR. ANNI MDCCLVII.

INDICIT

D. IO. ERNESTVS HEBENSTREIT

THERAPIAE P. P. FAC. MED. DEC. ACAD. DECEMVIR

COLLEG. MAI. PRINC. COLLEG.

CONSTITVTVS AD HVNC ACTVM

PROCANCELLARIVS.

Aetii Amideni Anekdoton liber IX, caput XXVI by Professor Hebenstreit (1757).

8

Genealogy of Laurentius Claussen

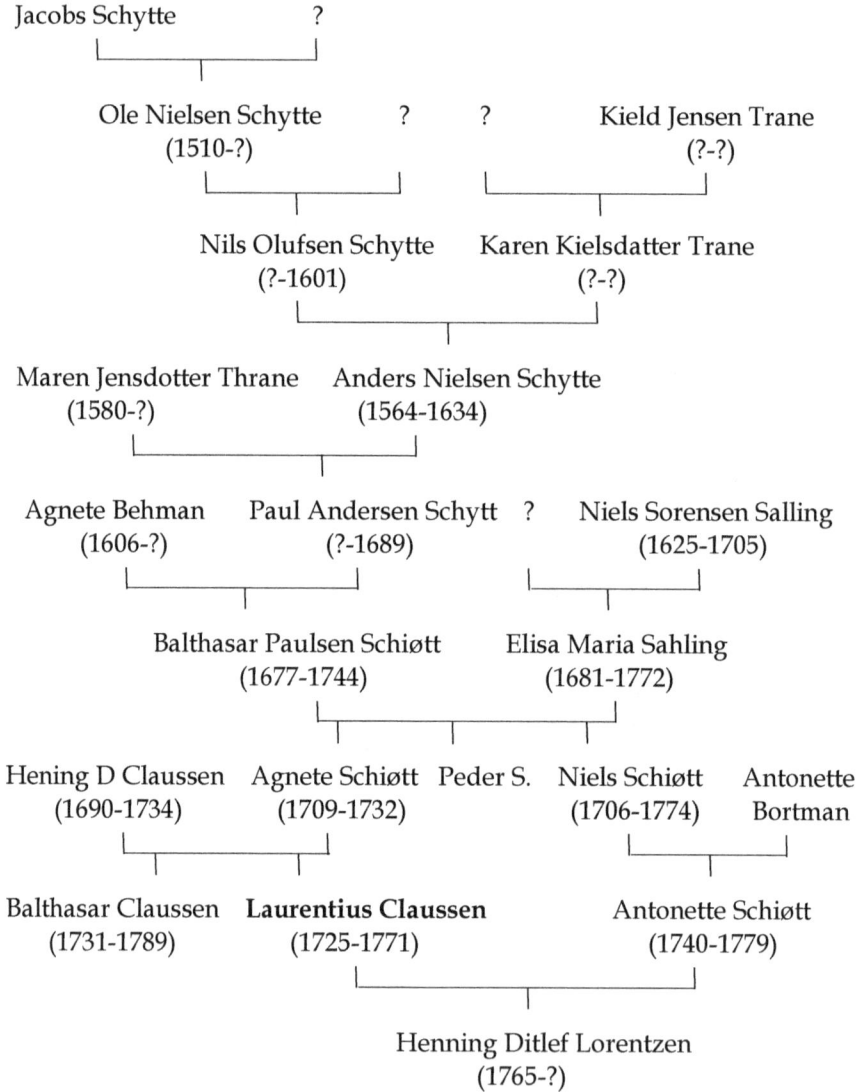

Jacobs Schytte   ?

Ole Nielsen Schytte   ?   ?   Kield Jensen Trane
(1510-?)   (?-?)

Nils Olufsen Schytte   Karen Kielsdatter Trane
(?-1601)   (?-?)

Maren Jensdotter Thrane   Anders Nielsen Schytte
(1580-?)   (1564-1634)

Agnete Behman   Paul Andersen Schytt   ?   Niels Sorensen Salling
(1606-?)   (?-1689)   (1625-1705)

Balthasar Paulsen Schiøtt   Elisa Maria Sahling
(1677-1744)   (1681-1772)

Hening D Claussen   Agnete Schiøtt   Peder S.   Niels Schiøtt   Antonette
(1690-1734)   (1709-1732)   (1706-1774)   Bortman

Balthasar Claussen   **Laurentius Claussen**   Antonette Schiøtt
(1731-1789)   (1725-1771)   (1740-1779)

Henning Ditlef Lorentzen
(1765-?)

In 1754, a certain Laurentius Claussen published the medical dissertation *De Tremore*. This was actually written by the contemporary and almost homonymous physician Laurentius Christianus Claussen who was actually born on Feb 28, 1730, in Glückstadt (or Tychopolis) in the Duchy of Holstein, then under Danish rule (today, in Germany). His father was the physician Michael Christian Claussen. This dissertation was presented in the University of Jena.

Q. D. B. V.

DISSERTATIO MEDICA INAVGVRALIS

D E

TREMORE

Q V A M

RECTORE ACADEMIAE MAGNIFICENTISSIMO
SERENISSIMO PRINCIPE AC DOMINO
D O M I N O
ERN. AVG. CONSTANTINO
DVCE SAXONIAE IVLIACI CLIVIAE MONTIVM AN-
GARIAE ET WESTPHALIAE, RELIQVA

P R A E S I D E
GEORGIO ERH. HAMBERGERO
PHILOS. ET MED. D. PHYSIC. CHIMIAE ET PRAXEOS PROF. P. ORD.
SERENISSIMORVM SAXONIAE DVCVM STIRPIS ERNESTINAE CON-
SILIARIO AVLICO ACADEM. NATVRAE CVRIOSORVM SODALI
CONCILII ACAD. ASSESSORE MEDICO PROVINC. IENENS.
ET FACVLT. MEDICAE SENIORE

PRO GRADV DOCTORIS
IVRIBVS AC PRIVILEGIIS DOCTORALIBVS
RITE OBTINENDIS
DIE XVII. IVLII ANNI M DCC LIV.
CONTRA ERVDITORVM DVBIA DEFENDET
A V C T O R
LAVRENTIVS CHRISTIANVS CLAVSSEN
TYCHOPOLIS: HOLSATVS.

IENAE LITTERIS RITTERIANIS.

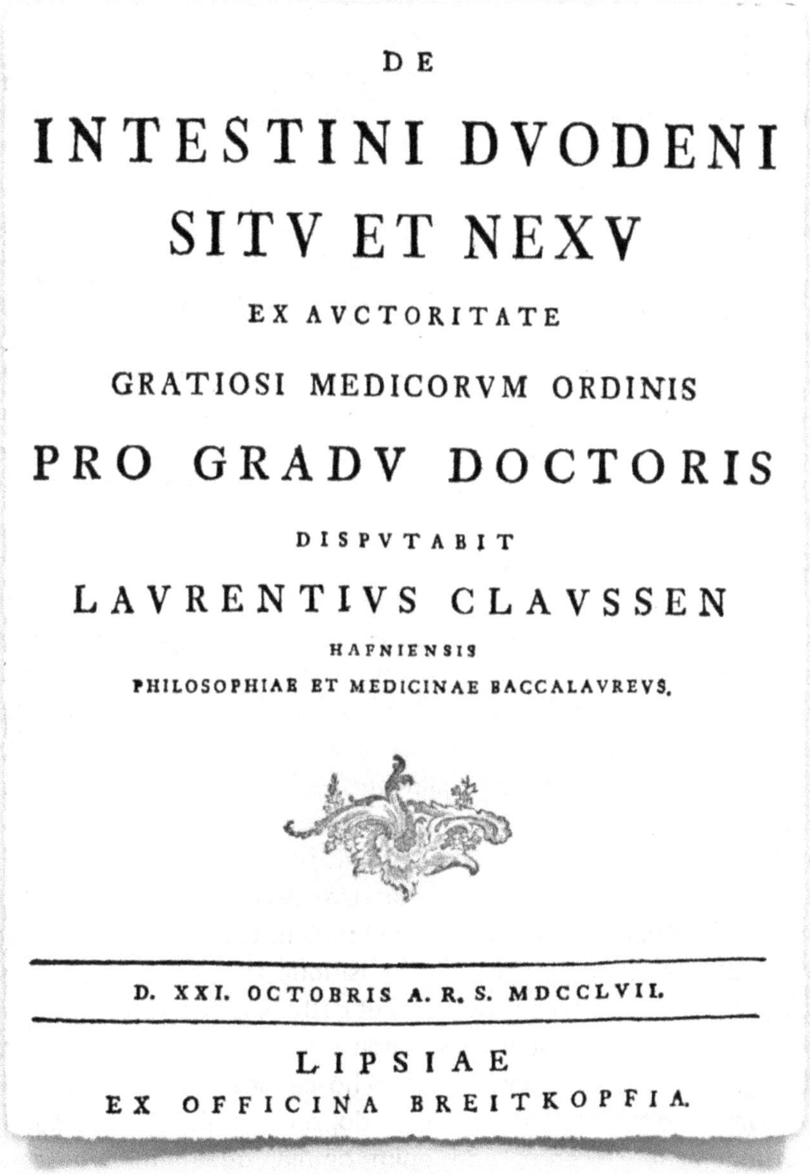

DE

INTESTINI DVODENI SITV ET NEXV

EX AVCTORITATE

GRATIOSI MEDICORVM ORDINIS

PRO GRADV DOCTORIS

DISPVTABIT

LAVRENTIVS CLAVSSEN

HAFNIENSIS

PHILOSOPHIAE ET MEDICINAE BACCALAVREVS.

D. XXI. OCTOBRIS A. R. S. MDCCLVII.

LIPSIAE

EX OFFICINA BREITKOPFIA.

Title page of *De Intestini Duodeni Situ et Nexu* (1757) by Laurentius Claussen.

12

PRAEFATIO

Cogitanti mihi, quamnam potissimum materiam dissertationi aptam eligerem, venerunt quidem in mentem varia, quae exponi a me posse videbantur: ea tamen praecipue studiis meis accomodata erant, quae Physiologiae et Pathologiae capita ex anatomicis doctrinis aliquo pacto illustrarent. Cum enim in Amphitheatro urbis patriae anatomico, Illustris Crygeri, Fautoris atque Praeceptoris summo cultu devenerandi, institutione uterer, ad haec studia praecipue excitatus fui. Inde in schola anatomica, quae regiis auspiciis Berolini floret, praeeunte Excellentissimo atque Experientissimo Meckelio, his exercitiis assidue vacare, et corporis humani partes omniaque viscera saepius perlustrare, maximae mihi voluptati fuit. Hinc, quum illustrem Philuream studiorum ulterius perficiendorum gratia adiissem, Magnificum et Excellentissimum Ludwigium novum optimumque Praeceptorem nancisci, eidemque cadaverum sectiones instituenti, atque viscera accuratissime monstranti, eorumque situm exponenti adesse iteratis vicibus contigit. Quae dum simul cum iconibus a variis auctoribus depictis conferre gaudebam, nullum ex iis inveni, qui intestinum duodenum, cui penitius perscrutando incubuimus, satis accurate exprimere videretur. Ergo in consortium ducto Clarissimo mihique amicissimo Reichelio, AA. M, novam delineationem elaborare tentavimus, confectisque novis qualibuscunque iconum lineis atque iisdem iterum iterumque correctis aliquam tandem paulo accurationem, ut opinor, obtinuimus, quam, in doctrina de situ et nexu duodeni illustranda speravimus, saltem non inutilem futuram. Hanc itaque materiam specimini elegi, et, disquisitione anatomica praemissa, consectaria quaedam physiologica et pathologica eruere, atque ita varia, circa usum duodeni attendenda, illustrare suscepi. Dum vero benigno et aequo peritiorum iudicio hosce meos labores submitto, veniam me impetraturum fore confido, si non ubique expectationi eorundem satisfecerim, non enim omnia, quae in illo viscere observanda sunt, exponere, sed ea tantum, quae praecipue situm et nexum huius intestini concernunt, ita declarare mihi proposui, ut simul nonnulla proferrem, quae in eiusdem sanitate curanda aut

PREFACE

When choosing the most appropriate topic for a dissertation, I recall that several which I felt capable of discussing had come to mind. These were especially adapted to my studies, where the principles of physiology and pathology were in some way explained from anatomical teachings. Indeed, I was particularly encouraged for such studies in the anatomical amphitheater of my native city, where I would enjoy the instruction by the esteemed **Krüger**, Patron and Instructor of the highest respect. In the school of Anatomy, which flourishes under the royal auspices of Berlin and under the guidance of the excellent and very experienced **Meckel**, I regularly attended these exercises and my greatest pleasure was to often examine all the organs of the human body. Hence, I undertook further studies in the illustrious city of Leipzig where I obtained a new and excellent teacher, the noble and magnificent **Ludwig**, from whom I learned to dissect bodies, accurately demonstrate organs and expose their location many times over. At the same time, I enjoyed comparing the illustrations portrayed by various authors which, upon closer scrutiny, did not depict the intestine called *duodenum* with sufficient accuracy. Therefore, in the company of a great friend of mine, the illustrious **Reichel**, we attempted to work out a new drawing, and produced new sketches of images which were corrected over and over again until we obtained, I believe, more accurate ones, since we hoped to elucidate the knowledge about the position and the relations of the duodenum for future use. And so, I have chosen this particular topic, and after a discussion of its anatomy I will expose certain physiological and pathological consequences and undertake to illustrate various matters regarding the function of the duodenum. As I submit these labors of mine to the kind and fair judgement of the experts, I am confident that I will succeed, even if I have not satisfied all their expectations, for I did not propose to explain everything that can be observed in this organ, but only that which mainly concerns the position and relations of this intestine, and to clarify what I investigated and which I believed could help cure and

restituenda medico profutura crediderim: in quo exiguo specimine corrigendo atque expoliendo vires meas olim ulterius experiri forte sustineam. Tu modo, Lector benevole, conatibus fave et vale.

....

(1) **Crygeri** = Simon Krüger (1687-1760) was a Danish barber-surgeon who, with others, founded in 1736 the "Theatrum Anatomico-chirurgicum," a school for teaching anatomy and surgery in Copenhagen, which prospered for the next twenty-four years but was suppressed in 1772. Krüger was an excellent teacher, but wrote very little (i.e., a couple of anatomical writings).

(2) **Meckel** = Johann Friedrich Meckel the Elder (1724–1774) was a German anatomist born in Wetzlar. Meckel earned his medical doctorate from the University of Göttingen in 1748. Subsequently, he moved to Berlin where he worked as a prosector and taught classes on midwifery. In 1751 he became a professor of anatomy, botany and obstetrics. Meckel has a number of anatomical eponyms associated with his name such as *Meckel's space*, *Meckel's ganglion* and *Meckel's ligament*. His grandson Johann Friedrich Meckel "the Younger" (1781–1833) was also an anatomist (of *Meckel's diverticulum* fame). The elder Meckel's son, Philipp Friedrich Theodor Meckel (1755–1803) and another grandson, August Albrecht Meckel (1790–1829), were also anatomists.

restore it to health. With luck I may continue to further prove my strengths in this small sample which is to be corrected and improved. As for you, kind reader, think kindly of my efforts and farewell.

....

(3) **Ludwig** = Christian Gottlieb Ludwig (1709–1773) was a German physician and botanist born in Brieg, Silesia (now Brzeg, Poland). He was the father of physician/naturalist Christian Friedrich Ludwig (1757–1823) and of physician/scientist Christian L. Ludwig (1749–1784). From 1728 he studied medicine and botany at the University of Leipzig, but due to lack of funds was forced to discontinue his studies, thereafter taking a job as a botanist on an African expedition under the leadership of Johann Ernst Hebenstreit (1703–1757). In 1733 he resumed his studies, and from 1736 gave lectures at Leipzig. In 1737 he earned his doctorate, later becoming an associate professor of medicine (1740). At Leipzig he successively became a full professor of medicine (1747), pathology (1755), and therapy (1758). Ludwig is remembered for his correspondence with Carl Linnaeus, and it was Linnaeus who named the plant genus *Ludwigia* in his honor.

(4) **Reichel** = George Christian Reichel (1717-1771) was born in the city of Mulhausen, in Thuringia, Germany. He enrolled as a medical student in Leipzig in 1747, obtained his baccalaureate in 1754, and his medical degree in 1759. In 1767 he was nominated as professor of medicine at the University of Leipzig. Reichel published his medical thesis *De Epiphysium ab Ossium Diaphysii Diductione* in 1759, and followed this with a study on the microscopic anatomy of bones *De Ossium Ortu atque Structura* (1760). Reichel's talents as an artist did not go unnoticed and he was recognized by his peers for the instructive copperplates in his medical writings.

§. I.

"De necessaria situs et nexus partium corporis humani disquisitiones agitur"

In usu viscerum et organorum corporis humani disquirendo non tantum ad fabricam ex figura, magnitudine et proportione partium componentium declarandam, sed simul etiam ad situm et nexum respiciendum esse illico patet. Musculorum actio et vsus in membris varie dirigendis huius asserti veritatem egregie evincit: Nemo enim motuum voluntariorum rationem scite explicare poterit, nisi ad situm et ordinem musculorum congenerum, eumque in varia artuum directione variantem, respiciat, et ita actionem quandam plus minusque validam definiat. Similis certe attentio in consideratione reliquarum corporis partium, imprimis viscerum, adhibenda est. Horum quidem fabricam in evolutione minimarum partium magno studio detegere annisi fuerunt recentiores dissectores, usus vero eorundem in oeconomia animali non satis patescit, nisi et cuiuslibet visceris situm et nexum cum reliquis attente perquiramus, et Medicus quoque, in facienda arte constitutus, in signis laesionis viscerum pathologice examinandis, et in vera morborum sede detegenda hallucinari facile posset, nisi et has partium proprietates bene cognitas habeat. Haec de duodeno quoque valere quilibet perspiciet, qui maximas, quas hoc intestinum in digestione alimentorum habet, partes perpendit. Et quamvis hoc intestinum celeberrimorum dissectorum disquisitionibus satis declaratum fuerit, nonnulla tamen circa descensum et ascensum, nec non circa varios eius flexus accuratius describenda sunt, quae uberius exponenda nobis sumimus.

CHAPTER ONE

"On the much needed investigation of the location and connections of the organs in the human body"

Upon investigating the function of the viscera and organs of the human body one needs to consider not only their structure, revealed from the shape, size and proportion of their component parts, but also their position and relations. The action and function of muscular activity of the variably arranged extremities brilliantly demonstrates this. In fact, no one can skillfully explain the nature of voluntary movements unless one considers the location and order of corresponding muscles, including their various directions within different limbs, and thus determine their motion of variable strength. Similar attention should be applied to the consideration of other body organs, especially the viscera. Recent anatomists have certainly strived to detect with great zeal the structure of these small organs. Indeed, their function is not revealed sufficiently in the physiology of animals, unless one researches attentively the location and relation of every organ with each other. Physicians, too, appointed to the practice of the profession, when examining the signs of injury of diseased organs, and searching for the true seat of the disease, can easily be fooled unless they have a good knowledge of the properties of such organs. This is also valid for the duodenum, as anyone who examines the most important digestive organs, which includes this intestine, will perceive. Although the duodenum has been sufficiently discussed in the investigations by celebrated anatomists, several issues about its descending and ascending segments, as well as its various curvatures, need to be described more accurately, and these we take upon ourselves to fully explain.

§. II.

"Prosectores, qui in Duodenum inquisiverunt, indicantur"

Est vero Duodenum prima intestinorum pars (a), a ventriculi orificio dextro ad principium ieiuni usque extensis, et quamvis a dimensione duodecim pollicum nomen receperit (b), ideo tamen a reliquo intestinorum tractu non distinctum fuit ob peculiarem potius, quam obtinuit, sedem, attentionem dissectorum praecipue promeruit (c). Quoniam vero Vesalius intestinum duodenum recta quodammodo eousque descendere dixerat, ibi canalis intestinorum in anfractus orbesque primum convolui incipit (d), plurimi dissectores eum in hac descriptione sequuti sunt. Columbus tamen optime monuit (e), hoc intestinum, postquam spinam versus descendit, assurgere itertum, et postea demum gyris in ieiuno intestino continuandis initium dare, quam ipsam ideam postea Regn. De Graaf (f) icone quodammodo adumbravit. Exhibuit vero, ut in omnibus solet, Winslowius accuratam huius partis descriptionem (g), et Garengeot demonstrationibus Winslowianis edoctus meliorem quidem praecedentibus, non tamen ubique accuratam, iconem proposuit (h). Monrous porro et descriptione et icone (i) singulares duodeni flexus accuratissime definivit, cui tandem concisa quidem, sed egregia Halleri (k) descriptio et delineatio addenda est.

CHAPTER TWO

"Anatomists who have investigated the duodenum are presented"

The *duodenum* is indeed the first part of the intestines (**a**), extending from the right orifice of the stomach to the beginning of the *jejunum*, and it received its name from its length of twelve inches (**b**). Nevertheless, although not distinct from the rest of the intestinal tract, it deserves the special attention of the prosector on account of the rather particular position which it keeps (**c**). Ever since **Vesalius** said that the *duodenum* somehow always descends in a straight line until where the intestinal canal first begins to roll into loops and circles (**d**), many prosectors followed him in this description. **Colombo** rightfully warned, however (**e**), that this intestine, after descending towards the spine, will ascend again, and then eventually give rise to loops that continue as the *jejunum*, an idea which **Regnier de Graaf** (**f**) later, in fact, depicted in an illustration. **Winslow** presented, as he always does, an accurate description of these organs (**g**) and **Garengeot,** with the knowledge of the demonstrations of Winslow, offered an improved illustration compared to the preceding, and yet not everywhere accurate (**h**). **Monro**, in turn, accurately designed the curves of the the *duodenum* with a remarkable picture and description (**i**), which is actually cut off at the end. On the other hand, the description and drawing of **Haller** (**k**) are excellent additions.

20

This chapter carried several footnotes, which will be examined in detail one by one since they offer a window into the background knowledge upon which Claussen built his dissertation.

>>>>>>>>>>>>>>>>>>>>>>>>>>><<<<<<<<<<<<<<<<<<<<<<<<

a) "εκφυσις" vel "εκφυσις δωδεκαδάκτυλος" vel etiam δωδεκαδάκτυλον veteres appellabant, quoniam a reliquis intestinis singularibus proprietatibus distinctum, non ut verum intestinum, sed ut ventriculi processum considerarunt. Vid. Ioh. Gorraei Definitionum medicarum Lib. XXIV. Lut. Paris. 1544. fol. pag. 130.

a) *Ecphysis* or *Ecphysis Dodecadactylos,* or even *Dodecadactylon,* as it is called by the Ancients, which because of its singular characteristics is distinct from the rest of the intestines; and it was not considered a real intestine but an extension of the stomach. See Iohannis Gorra, *Definitionum Medicarum, Libri XXIIII*, Lutetiae Parisiorum, 1544, folio, page 130.

>>>>>>>>>>>>>>>>>>>>>>>>>>><<<<<<<<<<<<<<<<<<<<<<<<

It was in 1564 that Jean **de Gorris** (1505-1577), a Parisian physician, wrote the Greek-Latin medical dictionary *Definitionum Medicarum, Libri XXIIII*. Further editions were issued later in 1578 and 1601, and a collected works edited by his son in 1622. The page number "130" mentioned by Claussen would tell us which edition he was referring to. In fact, the entries for the terms *dodecadactylon* and *ecphysis* can be found in the following pages:

	dodecadactylon	*ecphysis*
1564:	91	98
1578:	120	130
1601:	120	130
1622:	171	184

Thus, Claussen was using either a 1578 or a 1601 edition — a moot point since none differed in the definitions rendered. The only difference is that the 1st edition wrote the terms as δωδεκαδάκτυλομ or *dodecadactylom,* and εκφυσιμ or *ecphysim.*

These were corrected in the later editions. In de Gorris's book, the entries for the *duodenum* were under the words *Dodecadactylon* and *Ecphysis*:

"δωδεκαδάκτυλον: Principium est intestinorum ab imo ventriculi orificio exoriens. Id sub sede ventriculi posteriore, secundum dextram spinae partem deorsum exporrigitur, sine ullo anfractu, quo det locum quibusdam partibus quas natura inter ventriculum & ieiunum intestinum collocavit: duodecim digitorum longitudinem ut plurimum aequat, ob idque δωδεκαδάκτυλον ab Herophilo primum appellatum est. Sunt qui ipsum intestini nomine non dignentur, sed εκφυσιν modo appellent, vel cum addito εκφυσιν δωδεκαδάκτυλον. In id meatus inseritur a vesicula bilem continente, per quem ea in ipsum expurgatur."

[*Dodecadactylon*: It is the beginning of the intestines arising from the lowest orifice of the stomach. It extends beneath the posterior aspect of the stomach, then backwards along the right side of the spine without any curving, and is situated in the space between the stomach and the jejunum amongst several organs. It is twelve inches long according to many, so that it was first called *dodecadactylon* by Herophilus. There are some who do not dignify it with the name of intestine, but only call it *ecphysin,* or *ecphysin dodecadactylon.* Within it is inserted the meatus from the gallbladder through which this empties itself into the duodenum.]

"εκφυσις: est intestinum a ventriculo primum. Id ventriculo coniunctum est, nec protinus in circunvolutiones reflectitur, prius quam ijs locum praebuerit quae in medio inter ventriculum & ieiunum spatio locari necesse erat. Itaque cum intestinorum propria videatur esse circumvolutio, quidam ipsam εκφυσιν non dignantur intestini nomine, sed vel simpliciter εκφυσιν, vel εκφυσιν δωδεκαδάκτυλον, vel

IO. GORRÆI

PARISIENSIS,

Definitionum Medicarum

LIBRI XXIIII.

literis Graecis distincti.

§

LVTETIÆ PARISIORVM,

APVD ANDREAM WECHELVM, SVB
PEGASO, IN VICO BELLOVACO.

M. D. LXIIII.

Cum Priuilegio Regis.

etiam absolute δωδεκαδάκτυλον appellant, quod duodecim fere digitorum longitudinem aequet. Sunt vero & qui non totum eum processum εκφυσιν appellandum censent, sed ipsius tantum orificium quo dependet a ventriculo. Caeterum ad hoc intestinum meatus defertur, qui bile ipsam ex hepate in ipsum dimittit."

[*Ecphysis*: It is the first intestine after the stomach. It is connected to the stomach and does not turn itself into loops right away until it completes its course which is necessarily situated between the stomach and the jejunum. Thus, while it appears to be its own intestinal circumvolution, some people did not deign to give this *ecphysin* the name of intestine, but they simply called it either *ecphysin*, or *ecphysin dodecadactylon*, or even just *dodecadactylon*, because it usually equals the length of twelve fingers. There are others who do not think that the entire process should be called *ecphysin*, but only the orifice that hangs from the stomach. In addition, one finds in this intestine the opening from which the bile itself is delivered from the liver.]

Ecphysis or *ecphysin* is a Greek term that means "extension" or "outgrowth." This would imply that the *duodenum* is a part of or an attribute of the stomach rather than a distinct organ.

Here Claussen refers only to the fact that the duodenum is the first segment of the intestines. Neither in this chapter nor anywhere else in his thesis does Claussen mention the Greek terms of *ecphysin* or *dodecadactylon*, terms which were originally conferred to the duodenum.

24

>>>>>>>>>>>>>>>>>>>>>>>>>>><<<<<<<<<<<<<<<<<<<<<<<<

b) Limites duodeni a dimensione incertos esse ostendit Ioh. Riolanus Fil. in Opp. Anatomicis Lut. Paris. (1466) 1649. fol. in anthropologia Lib. II. Cap. XIV. pag. 102. et Frid. Ruyschius in adversariis Dec. II. p. 45. falso Veteres hoc intestinum duodenum dixisse monet, cum potius digitale vel intestinum rectum brevissimum dici debuisset, et ipse hac in re is disquisitionibus seductus.

b) Jean Riolanus (the Younger) in his *Opera Anatomicae,* Lutetiae Parisiorum (1649), folio, *Anthropographiae,* book II, chapter 14, page 102, showed that the limits of the duodenum are uncertain from its measurement; and Frederik Ruysch in *Adversariorum,* decade 2, page 45, warned that the Ancients mistakenly called this intestine the *duodenum,* instead it ought to be called digital or very short straight intestine, and in this regard he was led astray by his own investigation.

>>>>>>>>>>>>>>>>>>>>>>>>>>><<<<<<<<<<<<<<<<<<<<<<<<

The French anatomist Jean **Riolan** (1580-1657) wrote *Anthropographiae* in 1618 which was incorporated into the anthology *Opera Anatomicae* from 1649 with some minor changes. In regard to the duodenum, Riolan said:

"Intestinum primum quod ab ima sede ventriculi nascitur, ecphysis dicitur, ab Herophilo duodenum, quoniam olim duodecim transversos digitos longum erat, vix hodie quatuor, interdum sex digitos aequat, nec mensuram antiquam deprehendes, nisi graciliorem & angustiorem Ventriculi partem a fundo inferne exporrectam, usque ad anfractuum principium addideris, quam saepe duodecim digitos aequare vidi."

[The first intestine arises from the lowest part of the stomach, and is called *ecphysis* or *duodenum* by Herophilus since in the past it was twelve finger-breadths long, though today it is scarcely four, sometimes six, fingers, and you will not detect the ancient measurement, unless one adds the narrower and thinner part of the stomach which protrudes at the bottom of its fundus all the way to the beginning of the curve, which I then frequently observed as equal to twelve fingers.]

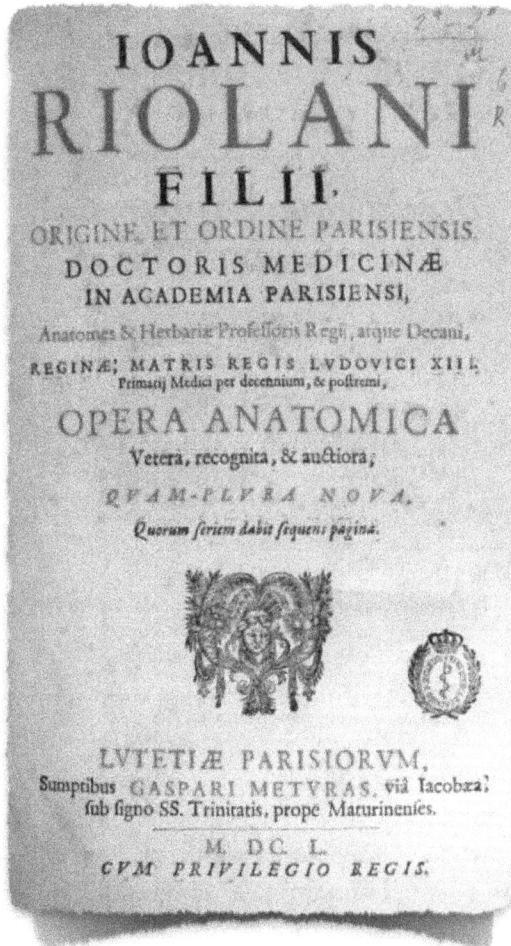

IOANNIS
RIOLANI
FILII.
ORIGINE ET ORDINE PARISIENSIS.
DOCTORIS MEDICINÆ
IN ACADEMIA PARISIENSI,
Anatomes & Herbariæ Profeſſoris Regii, atque Decani,
REGINÆ, MATRIS REGIS LVDOVICI XIII,
Primarij Medici per decennium, & poſtremi,
OPERA ANATOMICA
Vetera, recognita, & auctiora;
QVAM-PLVRA NOVA,
Quorum ſeriem dabit ſequens pagina.

LVTETIÆ PARISIORVM,
Sumptibus GASPARI METVRAS, viâ Iacobæâ,
ſub ſigno SS. Trinitatis, propè Maturinenſes.
M. DC. L.
CVM PRIVILEGIO REGIS.

"Nam ipse pylorus, primum est Intestinum, authore Celso, quinetiam Ruffus Ephesius & Pollux, primum Intestini exortum pylorum vocat, quod ex Galeno demonstrari posset, nisi Fuchsius in Paradoxis obstinate contrarium doceret. Semel mihi visum Duodenum, quod a pyloro usque ad inflexionem Intestinorum sex digitos transversos aequabat.

An anatomical dissection by Jean Riolan (engraving by R. van Persyn, 1649).

Istud autem Intestinum iuxta Hepar ad latus sinistrum contendit, & sub Pancreate iuxta Mesenterium investigandum.

Ubi intestina contorqueri incipiunt, ibi secundi Intestini, nempe Ieiuni principium sumitur, excipit meatum cholidocum, qui in confinium Duodeni & Ieiuni insertus, inter duas Intestini tunicas trium digitorum latitudine traducitur, perforatque internam tunicam: quidquid aliter sentiat Fuchsius in Paradoxis, & in suo lib. Anat. Ieiunum semper inanius apparere dicunt, quod non est absolute verum. Nam saepissime plenius deprehendi."

[For the *pylorus* itself is the first intestine, according to the author Celsus. Furthermore, Rufus of Ephesus and Pollux call the first of the intestines as *exortum pylorum*, which was pointed out by Galen, but Fuchs in his *Paradoxis* stubbornly taught the opposite. Once it seemed to me that the duodenum was six finger-breadths long from the pylorus to the intestinal curve. However, this intestine reaches near the left side of the liver and is observed under the pancreas next to the mesentery.

Where the intestine starts to form loops, that is, where the beginning of the second part of the intestine, namely the *jejunum*, commences, and receives the bile duct, which is inserted at the junction between the duodenum and the jejunum, travelling between the two layers of the three finger-wide intestine, and perforates through the inner layer; all this perceived by Fuchs in his *Paradox* and in his book on Anatomy. It is said that the *jejunum* always appears empty, which is not absolutely true, for it is more frequently detected full.]

Riolan's discussion revolves around the problematic length of the duodenum. Unfortunately, he believes that the *jejunum* begins at the papilla of Vater, thus foreshortening the duodenum itself. Not only does Riolan question the length of the duodenum, and thus its very name, but also disputes the basis of the name for the jejunum ("diet") since he found that it is full rather than empty at the time of dissection.

In 1736, Frederik **Ruysch** (1638-1731) wrote *Adversariorum Anatomico-Medico-Chirurgicorum*. On page 45, he said:

"Neque unquam vocassent in homine Intestinorum tenuium primum duodenum, si magna satis suppetisset incidendorum cadaverum copia. Sed vocavissent melius forte, certe magis apposite, intestinum digitale, vel intestinum rectum brevissimum. Multae utique partes male denominatae ex hac sola causa, quod ex brutis fuerit petita appellatio."

[Never would they have designated the first of the small intestines in humans as the *duodenum* if a large enough supply of corpses were available for dissection. But it would have been better, certainly more appropriate, if they had called it the finger-long intestine or the very short straight intestine. Certainly, many parts were ill-named for the sole reason that their name had originated from animals.]

Here Ruysch brings to the fore the argument that animals had a longer duodenum than humans. But this conclusion was based on his assumption that the duodenum ended at the point of entrance of the common bile duct into the duodenum.

FREDERICI RUYSCHII,

Anatomes & Botanices Professoris, Academiæ Cæsareae Curioforum Collegæ, nec non Regiæ Societatis Anglicanæ Membri,

ADVERSARIORUM

ANATOMICO- MEDICO-

CHIRURGICORUM

DECAS SECUNDA;

In quâ varia notatu digna recenfentur.

CUM FIGURIS ÆNEIS.

AMSTELÆDAMI,

Apud JANSSONIO - WAESBERGIOS.

M. DC. C. XXXVI.

The Anatomy Lesson of Dr. Frederik Ruysch by Jan van Neck (1683).

>>>>>>>>>>>>>>>>>>>>>>>>>>>><<<<<<<<<<<<<<<<<<<<<<<<<

c)   Vid. Dominici Santorini Observationes anatomicae, Venetiis
     1724. 4to maj. Cap. IX. §. 7. pag. 166.

c)   See Domenico Santorini *Observationes Anatomicae*, Venice
     (1724). quarto major, chapter 9, paragraph 7, page 166.

>>>>>>>>>>>>>>>>>>>>>>>>>>>><<<<<<<<<<<<<<<<<<<<<<<<<

In 1724, Giovanni Domenico **Santorini** (1681-1737) wrote the book
Observationes Anatomicae. Born in Venice, he earned his medical
doctorate in Pisa in 1701. From 1705 to 1728, Santorini performed
anatomical demonstrations in Venice, and in 1724 published his
detailed observations. Several anatomical structures were named
after him, notably the *Duct of Santorini* (the accessory duct of the
pancreas) and *Santorini's minor caruncle* (location of the opening
of the accessory pancreatic duct into the duodenum).

In his book, in chapter IX, page 166, Santorini begins his discussion of the *duodenum* with:

"Et circa Duodenum intestinum sic dictum quaedam sunt adjicienda minus hucusque animadversa, quae praeterire non praestat, eaque cum ad ejusdem positum, tum ad nonnulla intus contenta pertinet. Prius de positu dicam, caetera postmodum prosequuturus. Pylori orificio Duodenum continuari, nemo ignorant; illud vero modo per duos, tresve digitos eodem Pyloro superius esse, ac superincumbere, licet id perpetuum est, de Celebrioribus tamen Anatomicis haud quaquam intelligitur; quod porro flexuoso ductu leviter demissum inter felleam Cystim, & Pancreas sese conjicit; dein supra cavam, ac sub portarum venam paulum transversim incedens altero flexu ascendit iterum, circumundique adjunctis partibus firme alligatum, ut inde, ac ab eo positu minime dimoveri possit; sub Ventriculo locatur, ubi postremo flexu in acutum angulum conformatum deorsum versus convertitur, at laxe tum Mesenterio junctum in Jejunum abit; atque is Duodeni terminus est; Qui cum ab ejusdem fluxu, tum ab ejus nexu firmiore potius, quam a duodena digitorum mensura videtur constituendus."

[And regarding the so-called *duodenum*, there are certain things that need to be added which are insufficiently observed and cannot be neglected, and this pertains to its position as well as to some of the contents inside. Before I talk about its position, several other things need to be described in detail. As everyone knows, the orifice of the pylorus continues with the *duodenum*; this, in fact, rises for two or three finger-breadths above the pylorus and, even though it is continuous {with the remaining small bowel}, it is not given consideration by the most celebrated anatomists. Furthermore, this inserts itself as a slightly sunken flexible canal between the gallbladder and the pancreas, then it ascends again after a second flexure advancing a little transversely over the vena cava and under the portal vein, firmly tethered by neighboring organs, such that it can hardly be moved from this position. It is located beneath the stomach where it turns into the last flexure with a sharp angle towards the back, and then loosely joined to the mesentery it exits as the *jejunum*. And this is where the *duodenum* ends. It appears to be established by its course and its rather firm connections as much as from its length of twelve inches.]

Santorini describes in detail the three curvatures of the duodenum and stresses the importance of its position (*positu*) and relations (*nexu*).

>>>>>>>>>>>>>>>>>>>>>>>>>><<<<<<<<<<<<<<<<<<<<<<<<<

d)   Vid. Andreas Vesalius de fabrica corporis humani, Lib. V. p. 425. Fig. 12. Edit. Opp. Lug. Bat. 1725. fol. Tom. I.

d)   See Andreas Vesalius *De Fabrica Corporis Humani*, book 5, page 425, figure 12. Edition *Opera Omnia*, Leiden (1725), folio, tome I.

>>>>>>>>>>>>>>>>>>>>>>>>>><<<<<<<<<<<<<<<<<<<<<<<<<

In 1725, the monumental work of Andreas **Vesalius** (1514-1564), a Flemish physician (*aka* Andries van Wesel), was re-issued in the collected work called *Opera Omnia Anatomica & Chirurgica* by Boerhaave and Albini. On page 425 of book 5, chapter 5, Vesalius discussed the intestines:

"Primum quidem constituitur tota ea intestini pars, quae ventriculo substrata, ab inferiore ipsius orificio recta quodammodo eousque descendit, ubi intestinum in anfractus orbesque primum convolvi incipit. Ac de hujus intestini nomine aniles admodum contentiones apud illos reperies, qui reflectione neglecta, futiliter de voce altercantur. Sunt qui totam hanc partem quae duodenum digitorum longitudinem in viro ut plurimum aequat, non intestinum, sed εκφυσιν & exortum, seu processum, vocandum contendant. Alii, etiam si ipsius principium e ventriculo nascatur, intestinum tamen esse, ipsiusque orificium εκφυσιν, reliquum vero ipsius ductum, intestinum vocari affirmant, integros commentarios nugaci disceptatione implentes. Vocetur itaque nobis vulgato nomine tanquam succinctiore, duodenum, aut intestinum longitudinem duodecim digitorum aequans: siquidem id Graecis, δωδεκαδάκτυλον appellabatur."

ANDREAE VESALII

Invictissimi Caroli V. Imperatoris Medici

OPERA OMNIA

ANATOMICA

&

CHIRURGICA

Cura

HERMANNI BOERHAAVE

Medicinae, Botanices, Collegii Practici, & Chemiae in Academia Lugduno-Batava Professoris,

&

BERNHARDI SIEGFRIED ALBINI

Anatomes & Chirurgiae in eadem Academia Professoris.

TOMUS PRIMUS.

LUGDUNI BATAVORUM,

Apud { JOANNEM Du VIVIE, ET JOAN. & HERM. VERBEEK. } Bibliop.

MDCCXXV

[Indeed, {the *duodenum*} forms the first part of the entire intestines and lies distal to the stomach; it descends from the stomach's distal opening in a straight manner to the point where the intestines first begin to twist into coils and loops. Regarding the name of this intestine, you will find some disagreement among those who, because its coiling was overlooked, argue out loud and in vain. There are those who insist on calling this entire segment, which in man is equal to the length of twelve fingers, not an intestine but *ecphysin* & *exortum*, or *processus*. Others, however, even if its beginning arises from the stomach, say that it is an intestine and its orifice is the *ecphysin*, and they insist on calling the rest of its canal an intestine — a vigorous debate filled with

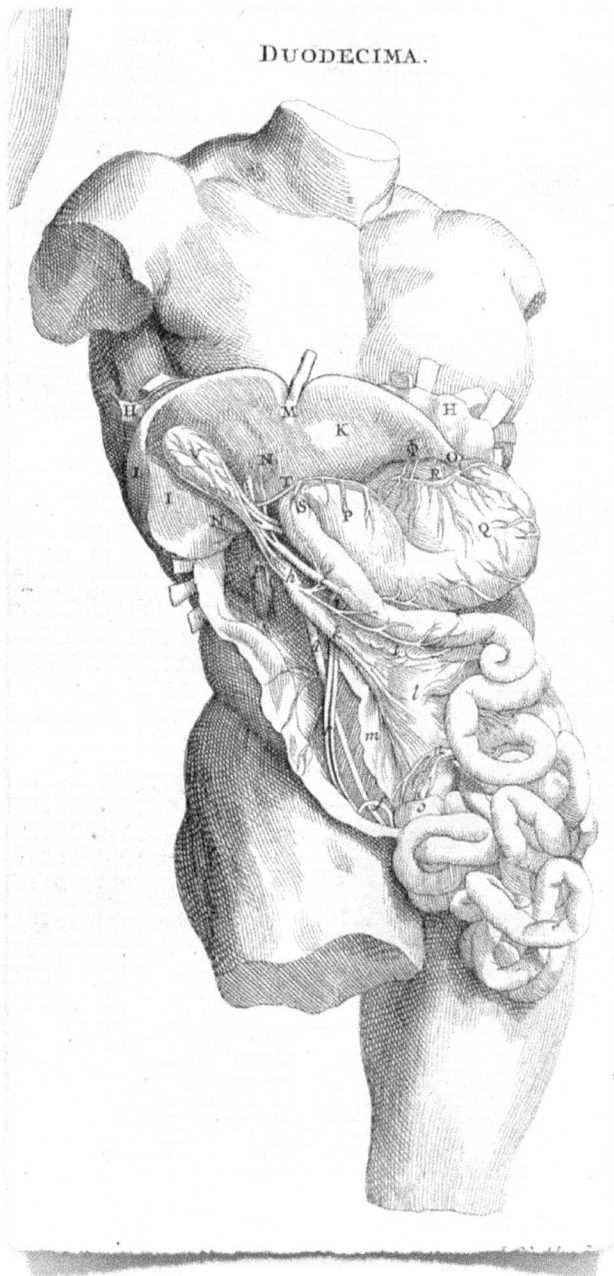

DUODECIMA.

frivolous commentaries. Let us then call it with the common short name of *duodenum,* or small intestine equal to the length of twelve inches: from the Greek, *dodecadactylon.*]

Claussen refers to figure 12 of Vesalius's book (see previous page) where one can see the duodenum descending in a straight course. The duodenum may seem somewhat curved but, as we will see below, Vesalius ends the duodenum at the insertion of the common bile duct. In fact, Vesalius had actually referenced his figure 7 in the footnote to the above paragraph, where he states:

"P 7 fig. ab I ad K."

[footnote P: figure 7, from I to K.]

This figure 7, in fact, delineates the extremities of the duodenum, from I to K, that is, from the pylorus up to the insertion of the bile duct — a fact that will not go unnoticed by many future anatomists.

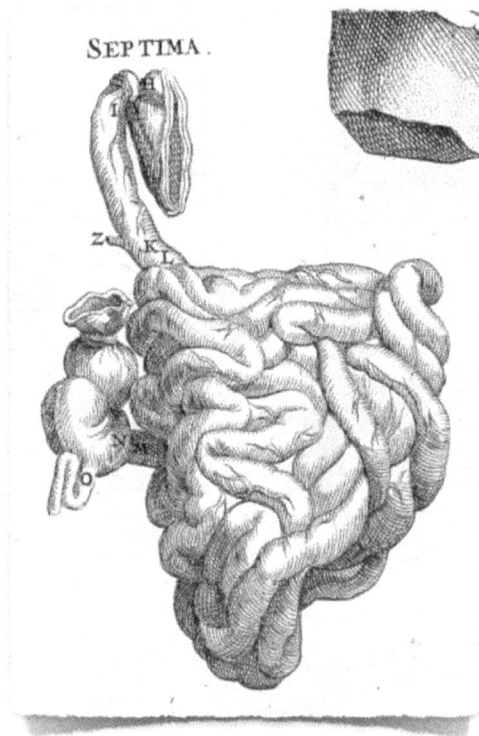

Legend: I = beginning of duodenum; K = end of duodenum; L = beginning of jejunum; Z = opening of the bile duct into the duodenum.

>>>>>>>>>>>>>>>>>>>>>>>>>><<<<<<<<<<<<<<<<<<<<<<<<

e) **De re anatomica Lib. XV. Venetiis 1559. fol. Lib. XI. de visceribus, pag. 228.**

e) *De re anatomica libri XV,* Venice (1559), folio, book XI. "De visceribus," page 228.

>>>>>>>>>>>>>>>>>>>>>>>>>><<<<<<<<<<<<<<<<<<<<<<<<

In 1559, the Italian surgeon and professor of anatomy Realdo **Colombo** (1516-1559) wrote the book *De Re Anatomica.* This was published in Venice, while a new edition was published in 1562 in Paris. On page 228, Colombo said of the *duodenum*:

> "Duodenum, ut ab hoc incipiamus; nam ventriculum subsequitur, ita appellatur, quoniam longitudinem eius duodecim digitorum apicibus metiri possumus: Graece dicitur dodecadactylos, dicitur & ianitor, portanarius, pyloros, aphysis: quae nomina nonnulli ad inferius orificium ventriculi transferunt. Duodenum post suum a ventriculo exortum, spinam versus descendit, quo postquam pervenit, assurgit, & gyris initium dat; ibique terminum habet, unde est origo ieiuno intestino, quod longum admodum est, sed duodeno tenuius."

> [We begin with the *duodenum,* for it follows the stomach, and is so called because its length can be measured as twelve fingers from its beginning. In Greek it is called *dodecadactylos*; it is also called *ianitor, portanarius, pylorus, aphysis {ecphysis}*: some of these names have been applied to the distal opening of the stomach. After its exit from the stomach, the duodenum descends towards the spinal column, and when it reaches this it ascends and gives origin to the loops; and there it ends where the origin of the *jejunum* is, which is longer but slender than the duodenum.]

Colombo is indeed one of the first anatomists to correctly notice that the duodenum does not only descend but thereafter *ascends* before becoming the jejunum.

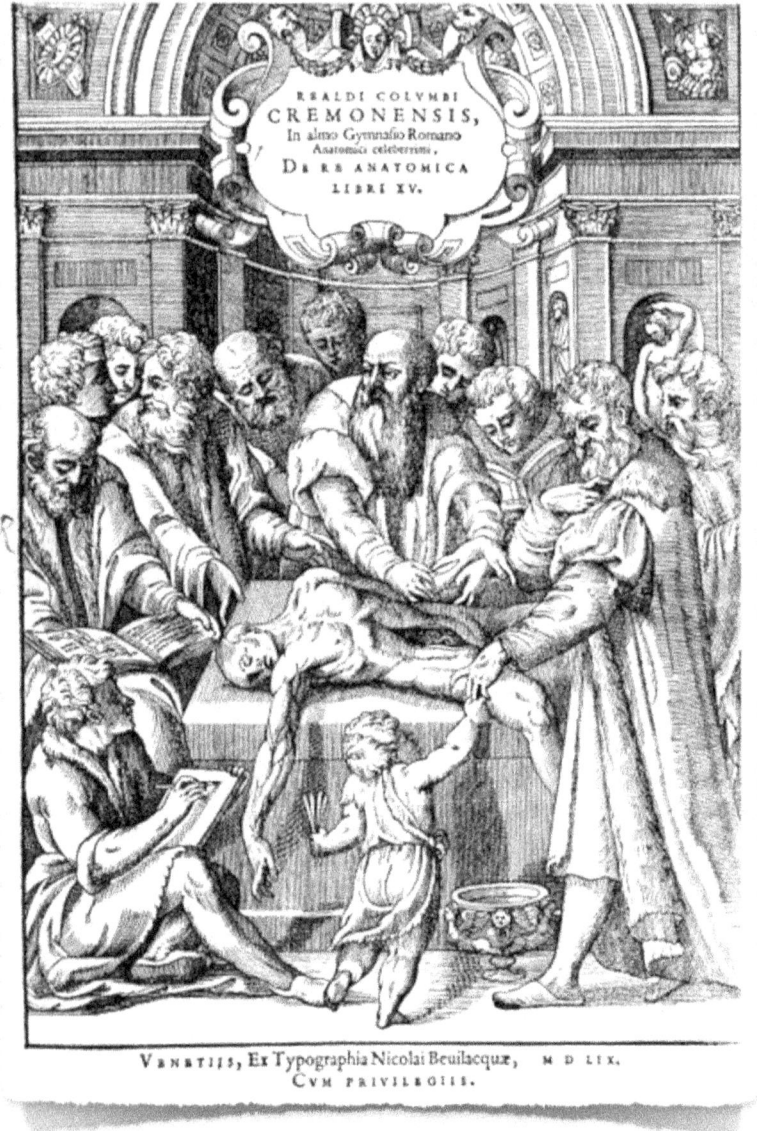

Frontispiece of the book *De Re Anatomica* by Realdo Colombo (1559).

>>>>>>>>>>>>>>>>>>>>>>>>><<<<<<<<<<<<<<<<<<<<<<<<<

f)   In tractatu de succi pancreatici natura et usu, Tab. I. in **Opp. Amstelodami 1705. 8vo.**

f)   In the treatise *De succi pancreatici natura et usu*, Figure I, in Opp. Amsterdam (1705), octavo.

>>>>>>>>>>>>>>>>>>>>>>>>><<<<<<<<<<<<<<<<<<<<<<<<<

In 1664, Reinier de **Graaf** (1641-1673) wrote *De Succi Pancreatici* which contained the illustration "Tabula I" depicting a very accurately C-shaped duodenum.

D = common bile duct; F = pancreas; H = stomach; I = pylorus; K = the beginning of the small intestines, called the duodenum; L = small bowel section under mesentery; Q = cystic duct; R = hepatic duct; S = opened duodenum.

38

Part "L" in this illustration is called **"Intestini tenuis pars sub Mesenterio mergens"** [Part of the small intestine diving under the mesentery]. This actually corresponds the the 3rd part of the duodenum, or inferior horizontal part, which then rises and crosses in front of the lumbar vertebrae and ends where the mesenteric artery "P" and vein "O" cross in front of it. This illustration is commendable for its proper depiction of the C-shape of the duodenum, something that had not been stressed by earlier authors.

>>>>>>>>>>>>>>>>>>>>>>>>>><<<<<<<<<<<<<<<<<<<<<<<

g) **Exposition anatomique de la structure du corps humain; traite du bas ventre, §. 104.**

g) *Exposition anatomique de la structure du corps humain;* "Traite du bas ventre," paragraph 104.

>>>>>>>>>>>>>>>>>>>>>>>>>><<<<<<<<<<<<<<<<<<<<<<<

In 1732, Jacques-Benigne **Winslow**, the Danish-born French anatomist, wrote *Exposition Anatomique de la Structure du Corps Humain*. This was translated into English in 1734 (*An Anatomical Exposition of the Structure of the Human Body*), into Italian in 1747 (*Esposizione Anatomica della Struttura del Corpo Umano del Winslow*), and into Latin in 1753 (*Expositio Anatomica Structurae Corporis Humani*). On page 510, paragraph 104 of the "Treatise of the Lower Abdomen," he wrote extensively of the *duodenum,* and we will reprint the relevant pararaphs no. 104 ('Name') and nos. 105-107 ('Situation and Connexion of the Duodenum'):

> "104. NOM. Cette premiere portion des Intestins Grêles a été ainsi appellée par rapport à la longueur de douze travers de doigt que les Modernes ne lui disputeront pas beaucoup, si l'on prend cette mesure avec les bouts des Doigts du sujet."

> "105. SITUATION. CONNEXION. Aussitôt que cet Intestin a pris sa naissance du Pylore, il fait d'abord une petite courbure en arriere, obliquement de haut en bas; ensuite il forme une seconde courbure vers le Rein droit, auquel il est plus ou'moins attaché; & de-là il passe devant l'Artere Renale,

EXPOSITION ANATOMIQUE
DE
LA STRUCTURE
DU
CORPS HUMAIN,

Par Jaques-Benigne WINSLOW, *de l'Academie Royale des Sciences, Docteur Regent de la Faculté de Medecine en l'Université de Paris, ancien Professeur en Anatomie & en Chirurgie de la même Faculté; Interprête du Roi en Langue Teutonique; & de la Societé Royale de Berlin.*

TOME TROISIEME.

A AMSTERDAM,
AUX DEPENS DE LA COMPAGNIE.
M. DCC. XXXII.

la Veine Renale & la Veine Cave, en remontant insensiblement de droite à gauche jusques devant l'Aorte & devant les dernieres Vertebres du Dos. Il continue sa route au-delà obliquement en devant, par un contour leger que l'on peut regarder comme une troisiéme courbure & comme l'extrémité du Duodenum.

106. Dans tout ce trajet le Duodenum est fortement attaché par des Replis du Peritoine, principalement par une Duplicature transversale qui donne origine au Mesocolon. Les deux Lames de cette Duplicature du Peritoine étant d'abord écartées l'une de l'autre & s'unissant un peu après, laissent naturellement entr'elles un espace triangulaire, dont le dedans est tapissé du Tissu Cellulaire.

107. C'est dans cet espace que le Duodenum est adherant par le Tissu Cellulaire aux parties que je viens de nommer, & qu'il est enfermé comme dans un Etui, de maniere que sans dissection on ne voit que ses deux extrémités, lesquelles sont encore cachées par le Colon & par les premieres circonvolutions de l'Intestin Jejunum."

[104. NAME. The first Portion of the small Intestine was called Duodenum, from the length ascribed to it by the Ancients, viz. the breadth of twelve Fingers; and the Moderns need not make petty objections about this length, if it is measured with the Ends of the Fingers of the Subject.

105. SITUATION. CONNECTION. This Intestine having arisen from the Pylorus, is immediately bent a little backward and obliquely downward, then it bends a second time toward the right Kidney, to which it is somewhat connected, and from thence passes in front of the Renal Artery and Vein, and the Vena Cava, ascending insensibly from right to left, till it gets in front of the Aorta and last Vertebrae of the Back. It continues its Course obliquely forward, by a gentle Turn, which may be reckoned a third Curve, and also the Extremity of the Duodenum.

106. Through this whole Course, the Duodenum is firmly bound down by Folds of the Peritoneum, especially by a transverse Duplicature which gives Origin to the Meso-

colon. The two Laminae of this Duplicature being at first separate, and soon after uniting, must leave a triangular Space between them, which is lined with a cellular Substance.

107. It is in this Space that the Duodenum adheres by means of the cellular Substance, to the organs already named, and the Intestine is contained therein, as in a Sheath (*Etui*), so that without Dissection, we can see nothing but its two Extremities, and even these are hidden by the Colon, and by the first Convolutions of the Jejunum.]

>>>>>>>>>>>>>>>>>>>>>>>>>><<<<<<<<<<<<<<<<<<<<<<<<

h) **Splanchnologie ou l'Anatomie des visceres, à Paris 1742. 8vo. Edit. sec. Tom. I. pag. 232.**

h) *Splanchnologie ou l'Anatomie des visceres*, Paris (1742), octavo, 2nd edition, Volume I, page 232.

>>>>>>>>>>>>>>>>>>>>>>>>>><<<<<<<<<<<<<<<<<<<<<<<<

In 1728, René Croissant de **Garengeot** wrote *Splanchnologie ou L'Anatomie des Visceres*. The 2nd edition was issued in 1742. On page 232, he gave the description to his 7th figure which highlighted the duodenum:

> **"Cette figure fait voir l'Estomac & le Duodenum soufflés, le foie, partie de la ratte, partie des reins, le pancreas, & le repli semi-lunaire du péritoine; le tout débarassé des autres parties, & dans sa situation naturelle."**

[This figure shows the Stomach & Duodenum distended with air, the liver, part of the spleen, part of the kidneys, the pancreas, and the semilunar fold of the peritoneum; all freed from the other organs, and in their natural position.]

SPLANCHNOLOGIE,

OU

L'ANATOMIE

DES

VISCERES;

AVEC des Figures originales tirées d'après les cadavres, suivie d'une Dissertation sur l'Origine de la Chirurgie.

Par RENE' CROISSANT de GARENGEOT, *Maître ès Arts & en Chirurgie, Démonstrateur Roïal d'Opérations, Conseiller Chirurgien Ordinaire du Roi en son Châtelet de Paris, de l'Académie Roïale de Chirurgie, & de la Société Roïale des Sciences de Londres.*

SECONDE EDITION,

Revûe, corrigée & augmentée par l'Auteur.

TOME I.

❀❀
❀

A PARIS,

Chez CHARLES OSMONT, Imprimeur de l'Académie Roïale de Chirurgie, rue S. Jacques à l'Olivier.

———————

M. DCC. XLII.

Avec Approbations & Privilége du Roi.

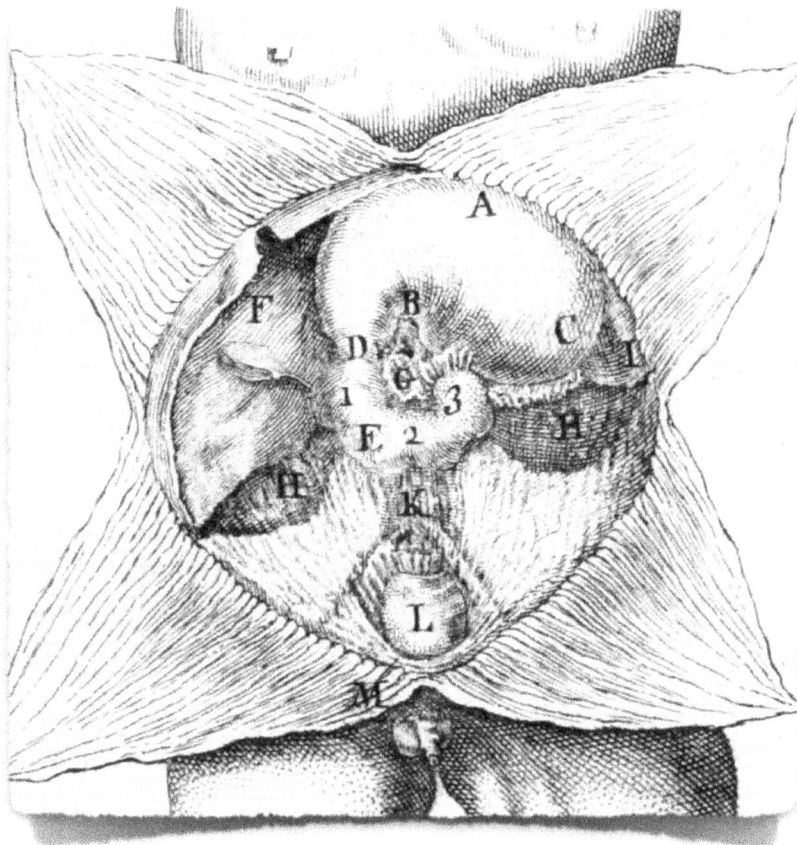

A = greater curvature of the stomach; B = lesser curvature of the stomach; C = the larger extremity of the stomach; D = pylorus; E = duodenum with its three curves 1, 2 and 3; F = liver; G = pancreas at the end of the 3rd curve of the duodenum; H = inferior pole of kidneys; I = anterior aspect of spleen; K = aorta; L = rectum.

This illustration by Garengeot is the first to depict the three curves of the duodenum and to demonstrate its course from the right to the left across the vertebral column.

44

>>>>>>>>>>>>>>>>>>>>>>>>>>>><<<<<<<<<<<<<<<<<<<<<<<<<

i) In Medical Essays and observations, Edinbourg 1738. Vol. IV.
    Obs. XI. pag. 65.

i) In *Medical Essays and Observations*, Edinburgh (1738), Volume
    IV, Observation XI, page 65.

>>>>>>>>>>>>>>>>>>>>>>>>>>>><<<<<<<<<<<<<<<<<<<<<<<<<

In 1738, Alexander **Monro** (the *first*), Professor of Anatomy in the
University of Edinburgh, wrote the paper "The Description and
Uses of the Intestinum Duodenum" in volume 4 of *Medical Essays
and Observations* (2nd edition). This is one of the most complete
discussions on the anatomy of the duodenum, most of which will
be here reprinted.

> "Anatomists have generally copied Vesalius's Description and
> Picture of the Intestinum Duodenum, which appeared to me
> very faulty. I caused Mr. Cooper to draw that Intestine in its
> natural Situation several years ago: Since that Time I have
> read two Authors, Santorini and Winslow, who have
> described this Gut more accurately than Vesalius; but neither
> of them having given any Figure of the Parts, and my
> Description differing considerably from theirs, as will appear
> upon comparing them; I resolved to send you this Paper, that
> the exact Situation of this Intestine might be more generally
> known, by which many Phaenomena in the animal Oeconomy
> and Diseases may be understood and explained.
>
> From the Pylorus, which is raised upwards and backwards
> from the Stomach, the Duodenum descends obliquely to the
> right Side, with the anterior Lamella of the Omentum fixed to
> its inferior Part; and the little Omentum, proceeding from the
> opposite Part, to connect it to the Liver. After this, the
> Duodenum is involved for about an Inch and a half, in a
> Doubling of the Omentum, and then enters into the
> Duplicature of the Mesocolon, where it can't be seen without
> dissecting away that fatty Membrane. It descends in this
> cellular Sheath, till it is almost contiguous to the great Sac of
> the Colon, which properly is the human Caecum. In this
> Descent the Colon lies before it; the biliary Duct, hepatic

artery and Nerve, vena portarum, and emulgent Vessels are behind it: The Liver, Gall-bladder and right Kidney are on its right Side, and the Pancreas is on the left. This Gut makes several Turns in this Progress; for it passes before the Vessels of the Liver. Immediately after, it bends backwards and to the right Side, till it approaches the right Kidney, and then turns forward, and a little to the left in its Course towards the great Sac of the Colon. The Duodenum then makes a considerable Curve to the left Side, where it is involved in a cellular Substance, which may be looked on as the common Root of the Mesentery and Mesocolon, through the Membrane of which it may be seen commonly, even in very fat Bodies, without any Dissection. In the concave left Side of this Curve, the thick Extremity of the larger Pancreas and the little Pancreas are lodged; the superior mesenteric Artery and Vein coming through the Notch between the larger and lesser Pancreas hang loose before the Gut here; and the Ductus communis Cholidochus, after passing behind the Gut a little higher, unites commonly with the pancreatic Duct, very little above the lowest Part of the Curve, and after passing obliquely through the Coats of the Gut, the two Ducts open by one common Orifice in the posterior Part of the Duodenum. After the Curve just now described, the Duodenum is involved in the Root of the Mesentery, and mounts obliquely within it towards the left Side, with the Vena Cava behind it; and after a Course of about four Inches there, rises forwards, to acquire a proper Mesentery, or to commence Jejunum, the Membranes of the Root of the Mesentery seeming to make a Ring at which the Gut comes out, though they are really continued on the Intestine, and forms its external membranous Coat.

That the Duodenum may be all exposed to View, without changing its natural Situation in a Body lying supine, it is necessary to cut through the great Arch of the Colon below the Bottom of the Stomach, and after turning the cut Extremity of the left Side over on the left short Ribs, to take hold of the other Extremity of the Colon; and having separated it with a Pair of Scissars from the Stomach and

Liver, taking away with it as much of the Omentum and Mesocolon as obstruct the View of the Duodenum and Pancreas, to lay it likewise on the right Loin. When the Colon is removed, observe where the Roots of the Mesentery and Mesocolon cover the Duodenum so much as to prevent your seeing its Course; at such Places cut these Membranes with a very sharp Scalpel, directing the Incisions according to the Length of the Gut, and then cautiously separate the Membranes to each Side, till all the Intestine is in view. Lastly, draw the small Guts gently down, raise the Liver, and suspend the Fundus of the Stomach as much as is necessary to allow a full View of the whole Course of the Duodenum.

Those who have ever dissected the human Body must be sensible how difficult it is to lay the Duodenum of an Adult all in View, without disturbing its Situation, and the Task of keeping all the Parts in the same fit Posture till a Painter delineates them, is still much greater: Therefore, though the preceding Description is taken from the adult Body, I chused to lay the Body of a Fetus, which I had preserved several Years in acidulated Spirit of Wine, before Mr. Cooper, to draw the Picture from, and afterwards I compared this Picture with several adult Bodies, to make sure of there being no essential Difference.

In *Tab*. I. *Fig*. 1. are represented:

. . . .

G. The *Pylorus* where the *Duodenum* begins, and the *little Omentum* connects it to the Liver. From this to H it is covered by the *Omentum*. Between H and I this Gut is lodged in the cellular Substance of the *Mesocolon*, thence to K it is covered by the common Root of the *Mesocolon* and *Mesentery*. It runs involved in the *Mesentery* to L, where there is an Appearance of a Ring; but instead of being turned down afterwards, as here represented, because of the Guts being drawn so much down to have a full View of the *Duodenum*, this Gut makes the Curvature delineated in *Fig*. 2.

. . . .

From the description of the Duodenum it must appear,

1. That since it is involved in the cellular fatty Substance of the Omentum, Mesocolon and Mesentery, without having the firm external Membrane braced upon it as the other Guts have; it must therefore more easily yield to any distending Force: And having the whole Substance thrown into the Stomach with the Bile and pancreatic Juice poured into it, it must receive more than any other Intestine; and then whatever enters it, must go out with some Difficulty, because its Extremity next to the Jejunum, is fix'd in a Course almost perpendicular upwards. So that upon the whole, it is no Wonder that this Intestine is frequently found of so much larger Diameter than the other Guts, as to be called Ventriculus Succenturiatus by several Authors.

2. The ascending Course of the Extremity of this Gut, and the Influx of the Bile and pancreatic Liquor into the most depending Part of it, where the Food must make the longest Stop, are wisely contrived, both for the more easy Influx of these Liquors, and for a sufficient Quantity of them being mixed with the Food, to perform well the necessary Offices for which they are designed in Digestion.

3. A pendulous Intestine here would, in our erect Posture, have drawn the Stomach out of its due Situation, and might have twisted or overstretched the biliary and pancreatic Ducts, so as to have stopped the Course of the Liquors in them; and therefore it is so firmly tied down in its whole Course, that it cannot change its Situation.

The Duodenum of Brutes is likewise placed in such a manner as to answer the same useful Purposes, though in many of them this Gut would appear to one who does not consider the different Postures and way of Life of Animals, to be situated in an opposite manner to the human Body."

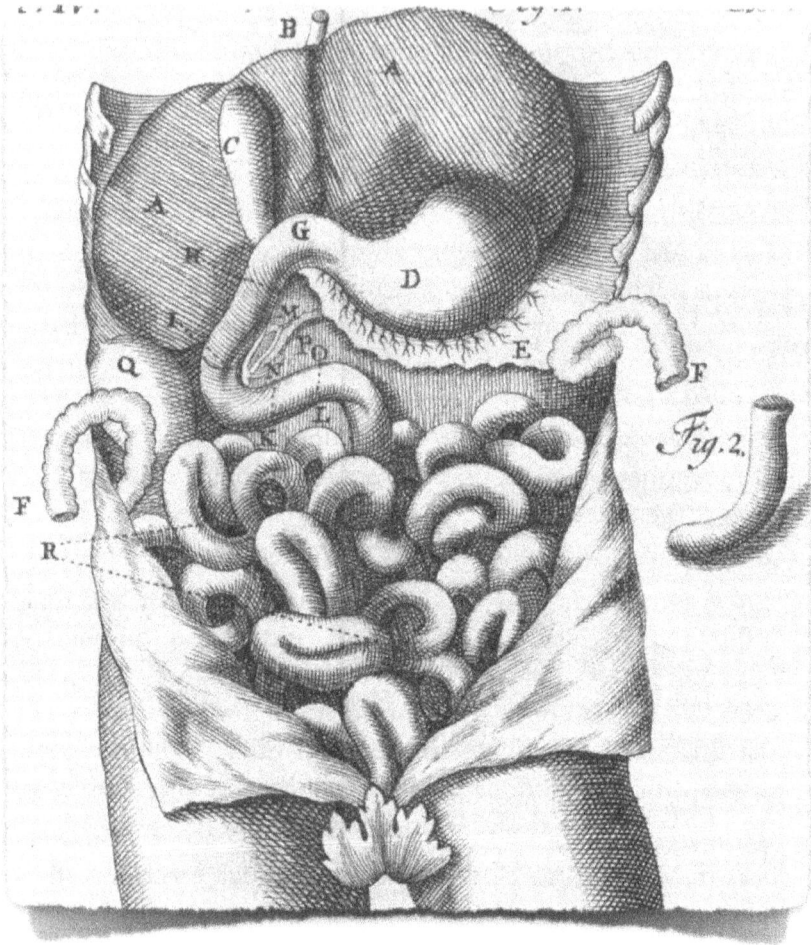

A = liver; B = umbilical vein; C = gallbladder; D = stomach; E = omentum; F = "arch" of the colon {transverse colon}; G = pylorus; G to L = duodenum, divided in the following sections:

> G to H = duodenum covered by omentum;
>
> H to I = duodenum covered by mesocolon;
>
> I to K = duodenum covered by root of mesentery;
>
> K to L = duodenum involved in mesentery;

M = greater pancreas; N = lesser pancreas; P = common bile duct; Q = right kidney; Fig. 2 = upward curve of duodenum at its distal end.

As noted by Claussen, Monro's anatomical representation of the duodenum with its curves and topographical position is, indeed, the most accurate to date.

>>>>>>>>>>>>>>>>>>>>>>>>>>>><<<<<<<<<<<<<<<<<<<<<<

k) In Boerhaavii praelectionibus academicis, Vol. I. Göttingae 1740. 8vo P 96. pag. 408. Not. I. Conf. Icon Anat. Part. C. H. Fasc. secundo Arteriae coeliacae, Tab. II. lit. H et I. Götting. 1745. fol.

k) Boerhaave's *Praelectiones Academicae*, Volume I, Gottingae (1740), octavo, paragraph 96, page 408, note 1; & compare with *Iconum Anatomicarum Partium Corporis Humani*, Fasciculus II, "Arteriae Coeliacae," Tabula II, letters H and I, Gottingae (1745), folio.

>>>>>>>>>>>>>>>>>>>>>>>>>>>><<<<<<<<<<<<<<<<<<<<<<

Here, Claussen references two writings of the Swiss anatomist Albrecht von **Haller**. One is the *Praelectiones Academicae* from 1740 which contains the lectures of Boerhaave with footnotes and commentaries by his pupil Haller. Within Boerhaave's text, wherein he speaks of the straightness of the *duodenum*, Claussen specifically highlights Haller's footnote "1" to a paragraph in the body text, both of which are reproduced here. Boerhaave said:

"**Rectitudo: Hoc solum Intestinum, & initium Jejuni recta (1) sunt, reliqua omnia in mirabiles, & vix describendos gyros implicantur.**"

[*Straightness*: Only this intestine and the beginning of the jejunum are straight (1), all the others are involved in incredible and hard to describe gyrations.]

Haller then added his commentary to Boerhaave's statement:

"**[1] Non possum tueri sententiam PRAECEPTORIS, quae neque tum satis firma est, quando Duodeni finis ad insertionem ductus Choledochi ponitur; Neque terreor auctoritate RUYSCHII, qui Duodenum 'intestinum Rectum brevissimum' vocat, Decad. Advers. II. p. 45. Nempe Duodenum ex Pyloro primo leniter sursum (EUST. T. X. f. 3.**

SANTORIN. c. IX. P. VII. MONROO Edimb. Ess. IV. p. 66)
deinde magis retrorsum (WINSL. IV. 105) & paulum deorsum
pergit, ita ut duplici, vel triplici alternatione iter in universum
transversum emetiatur ad usque vesiculam fere fellis. Tunc
vero descendit ad angulum rectum, & prius sinisterius, dein
dexterius se in Mesocoli duplicationem insinuat, atque ita per
inferiorem laminam pellucet. Sed brevi postea inflectitur
sinistrorium in ea ipsa duplicatione Mesocoli, & simul
adsendit modice, atque ante Aortam comite vena Renali
dextra pervenit ad communem Mesocoli transversi, &
Meseneterii radicem pone Vasa magna Mesenterii, tumque
circa eam radicem conflexum descendit emergendo in
abdomen inferius, estque Jejunum. Confer icones GRAAFII de
Succo Pancr. EUSTACHII, & MONROI. Duodeni vero
adsurgens origo postquam flatu distentum est, fit
descendens WINSL. l. c. confer icones GARENG. T. VII & X, f.
1."

[(1) I cannot support the opinion of the Instructor
{**Boerhaave**}, who is not to be relied upon much when he
places the insertion of the bile duct at the end of the
duodenum, nor am I frightened by the authority of **Ruysch**,
who calls the duodenum "the very short straight intestine,"
(in *Adversariorum,* 2nd decade, page 45). Of course, the
duodenum first rises up from the pylorus (in **Eustachius**,
volume 10, figure 3; **Santorini**, chapter 9, paragraph 7; in
Monro, *Essays of Edinburgh,* volume 4, page 66) and then
gently turns back (in **Winslow**, volume 4, page 105) and a
little downward, so that it imparts a double or triple shift in
its entire transverse course all the way to the gallbladder.
Afterwards, it descends at a right angle, first to the left and
then to the right, and insinuates itself into the root of the
mesocolon, and subsequently it appears through the lower
layer. But shortly afterwards it turns left within the root of
the mesocolon and, as it ascends a little, passes in front of
the aorta accompanied by the right renal vein and reaches
the common transverse mesocolon, behind the root of the
mesentery and the greater mesenteric vessels, and then
descends with a turn near this root to emerge into the lower

abdomen, where it becomes the *jejunum*. Refer to the illustrations of **Graaf** in *De Succo Pancreatico*, of **Eustachius** and of **Monro**. The ascending origin of the duodenum after it is distended with gas becomes descending; **Winslow** in cited work; consult the illustrations of **Garengeot**, volume 7 and 10, figure 1.]

The second reference of Claussen's footnote (k) is an illustration in the article by Albrecht Haller on the celiac artery and its branches in the 2nd fascicle of Haller's *Iconum Anatomicarum Partium Corporis Humani* from 1745 (see next page).

This reference only points to Haller's figure 2, and specifically to items:

H. Duodeni portio descendens.
I. Duodeni portio transversa altera.

H = descending part of the duodenum;
I = transverse or second part of the duodenum.

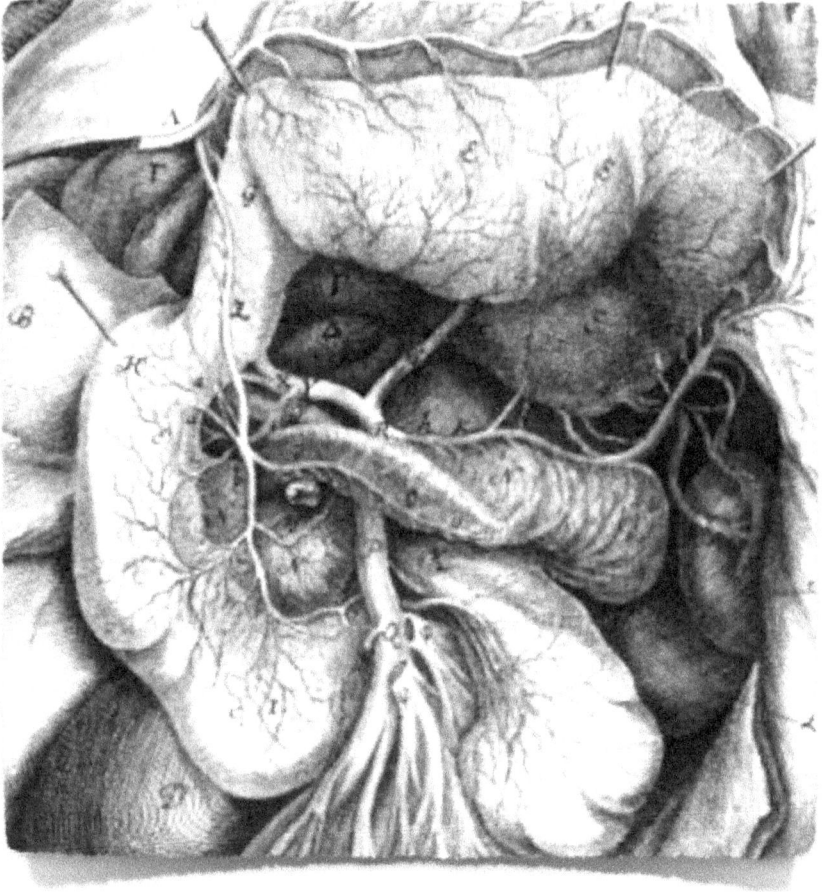

Iconum Anatomicarum Partium Corporis Humani, Fasciculus II, "Arteriae Coeliacae," Tabula II, letters H and I, Gottingae (1745)

"H" is near the pin on the left and indicates the descending part of the duodenum.

"I" is at the bottom left and indicates the transverse or second part of the duodenum.

§. III.

"Duodeni disquisitionem et delineationem difficile est ostenditur"

Ex accurata anatomicorum nunc citatorum perlustratione, in descriptionis et iconum comparatione aliquem inter eos deprehendi dissensum, cognoscimus, qua in re diversa quidem, obesa et macilenta, infantilia et adulta corpora differentiam passim notandam efficiunt, praecipua vero causa in difficili huius partis intestinorum disquisitione posita esse videtur (I), quam quidem, antequam descriptionem nostram exhibemus, paucis perpendere e re nostra esse ducimus. Quodsi enim intestinum nostrum in nexu cum reliquis partibus vicinis examinare velimus, quae vera est investigationis anatomicae ratio, illud adeo absconditum deprehendimus, ut non nisi unam vel alteram eius partem prominere videamus, ideoque non nisi successive omentum, ventriculum, hepar et colon reclinando, totum decursum emetiri queamus. Quodsi vero viscera reliqua ex situ suo dimoveantur, tunc vera relatio partium inter se, quae in vero corporis statu usus genuini definitionem efficit, attendi vix poterit. Porro quoque non ad viscera vicina tantum, sed etiam ad pinguedinem appositam et productiones peritonaei, quae partes vicinas connectunt, respiciendum est: si enim haec intacta reiinquuntur, nihil ferme apparet, si vero laxa haec vincula resecantur, quo omnis duodeni ambitus in conspectum prodeat, flaccidum flatuque distentum intestinum extra situm ducitur, et longe aliam formam sistit, quam in statu corporis vivo unquam assumere possumus. Maxima tandem oboritur difficultas, si delineatio apta, sine qua tamen omnis descriptio manca est, adornari debet. Quemcunque enim situm eligat delineator, totum huius intestini ambitum uno intuitu rite perspicere nunquam poterit; si enim ad cadaveris pedes collocatus delineationem suscipit, multas quidem detegere, et in iconem coniicere potest conditiones, situm huius visceris illustrantes, flexum tamen praecipuum, qui sub hepate a latere vesiculae felleae versus renem dextrum deducitur, nunquam apte

CHAPTER THREE

"The investigation and drawing of the duodenum is difficult to demonstrate"

From a careful review of the anatomists just cited, we realize that there are signs of disagreement which can be found when comparing their descriptions and illustrations. In fact, the bodies of the obese and thin, or infants and adults, are frequently very different in various ways, and a special contention appears to be placed in particular upon the difficult investigation of the duodenum (1). Thus, before we present our own description, we believe that it is in our interest to carefully examine a few of these. For if indeed we wish to examine the duodenum in relation to the rest of the neighboring organs, which is the true reason for the anatomical investigation, we discover that it is so hidden, that only one or another of its parts is visible, and for that reason only by bending back the omentum, the stomach, the liver and the colon are we able to measure its entire course. But if the rest of the viscera are moved away from their natural position, then we will not be able to consider the true relationship of these organs to each other, which would give an authentic explanation of their natural function in a body in standing position. Furthermore, one must consider not only the neighboring organs but also the nearby fat and peritoneal membranes, which are capable of connecting organs together. If these are left untouched, virtually nothing is visible. However, if these bonds are cut loose, then every area of the duodenum comes into view. This flaccid intestine, when distended by gas, is carried far from its true position, and it presents a very different shape than it actually assumes in the living body. Finally, the greatest difficulty arises if a suitable drawing must be prepared, without which any description is incomplete. Whatever position one chooses to sketch from, one will never be able to discern the entire length of this intestine from only one viewpoint. If one takes up the drawing while situated at the feet of the cadaver, much can be certainly discovered, and one can include in the drawing the features that elucidate the location

cognoscet. Si vero a dextro cadaveris latere positus hunc ipsum tractum attendit, tunc finis nostri intestini versus principia ieiuni excurrens, vel obscure tantum patet, vel in situ obliquo conspicitur.

of this organ. However, one may never properly recognize the particular flexure of the duodenum as it travels under the liver from the side of the gallbladder towards the right kidney. If, however, the artist is situated on the right side of the cadaver and directs his attention to this very tract, then the end of the duodenum as it reaches the beginning of the *jejunum* will either appear indistinct or is seen in an obique position.

....

This chapter included the following footnote:

\>>>>>>>>>>>>>>>>>>>>>>>>>>>><<<<<<<<<<<<<<<<<<<<<<<<

l)   **vid. MONROUS loc. cit. p. 69.**

l)   see Monro, cited work, page 69.

\>>>>>>>>>>>>>>>>>>>>>>>>>>>><<<<<<<<<<<<<<<<<<<<<<<<

This footnote refers us to the article "The Description and Uses of the Intestinum Duodenum" written in 1738 by Alexander **Monro** (the *first*) in the 2nd edition of his *Medical Essays and Observations*. In fact, on page 69, Monro said:

> **"Those who have ever dissected the human Body must be sensible how difficult it is to lay the Duodenum of an Adult all in View, without disturbing its Situation, . . . "**

§. IV.

"Optima duodenum disquirendi et delineandi ratio proponitur"

Omnibus vero his, in disquisitione oriundis, difficultatibus, quantum fieri potuit, occurrere, et ita veram situs et nexus ideam sequentem in modum attingere tentavimus. Ligaturam nimirum ieiuno ad pollicis circiter distantiam ab eius origine iniecimus, quo ventriculus et duodenum, paulo magis, quam reliqua intestina, flatu distendi queant, nimiam autem distensionem caute evitavimus. Paulo post, disquisitione repetita, omnem eius tractum ita consideravimus, uti reclinatis quodammodo, minime tamen resectis, visceribus vicinis in conspectum prodit. Cum hoc examen non sufficeret, ieiuno prope ligaturam praescisso omnem reliquum tenuium intestinorum canalem ita removimus, ut mesenterii radix, sive pars posterior vertebris annexa, duodeno incumbens integra relinqueretur. Zonam coli autem cum mesocolo in media sui parte ita divisimus, ut sinistra pars cum mesocolo adhaerente sursum reclinari, dextra vero, resecta mesocoli parte, quo duodeni ambitus pateat, magis dextrorsum reduci posset. Ventriculus flatu modice distentus paululum quidem, neque tamen nimium versus sternum elevandus est, quo curvatura magna sursum magis, quam antrorsum dirigatur. Hepatis quoque ratio habenda est, hoc enim viscus in cadavere, supino situ collocato, inprimis, si viscera reliqua e sede dimota sunt, non nihil deorsum et introrsum cedit, hinc hepar margini suo anteriori atque acuto ad costas dextri lateris quodam modo affigere, situmque eius, quantum fieri potest, naturalem praestare convenit. His in corpore dissecto rite dispositis, et intestino nostro flatu modice expanso, omnis ambitus oculo huc illucque flexo rite examinari, et, ut decet, describi potest. Delineatio pefecta, quae scilicet uno intuitu totum nostrum viscus oculis sistat, vix exaranda est. Quamvis enim, delineatore ad pedes collocato, et cadaveris superiore parte praecipue thorace aliquantum elevato, multa conspiciantur, et quam maxime progressus transversalis inferior a rene ad mesocolon et ad ieiuni principium egregie pateat, tamen et flexus principii duodeni a pyloro deductus, et lateralis sub

CHAPTER FOUR

"The best method to investigate and draw the duodenum is proposed"

We will certainly attempt to resolve as much as possible all the problems which came up in the investigation, and thus deal with the true notion of the position and relationships of the duodenum in the following manner. We placed a tie on the jejunum at a distance of about an inch from its origin, so that the stomach and duodenum could be distended by gas, a little more than the rest of the intestines, but we cautiously avoided any excessive distention. Shortly after, continuing the examination, the whole tract is inspected as it comes into view when the nearby organs are bent back a little, but minimally resected. Since this examination is not enough, with the jejunum cut away near the tie, we remove the entire remaining small intestinal canal, so that the root of the mesentery, or its posterior part attached to the vertebrae, is left intact lying on the duodenum. We then divide the area of the colon with its mesocolon in its midportion. When the mesocolon is so divided, the left side with the attached mesocolon is folded up. On the right, however, the resected part of the mesocolon, which exposes the area of duodenum, can be folded back more to the right. The stomach being moderately distended by gas should not be lifted too much towards the sternum where the greater curvature would be directed more cranially rather than forwards. Also, attention should be paid to the liver. In fact, in the corpse placed in the supine position, this organ does move a bit downward and inward, especially if the remaining organs are moved from their position; hence the anterior and sharp margin of the liver in some way should be fastened to the right lateral ribs, in its natural position, as much as possible. Having these been properly arranged in the dissected body, and the duodenum moderately expanded by gas, one can properly examine with the naked eye the entire area as well as its flexures, and, as it is suitable, describe it. The perfect drawing, in which the entire duodenum is visible to an attentive observer, is

hepatis concava parte ad renem usque continuatus descensus non satis ostendi potest, quam ob causam alia icon conficienda erat, quae delineatore a dextris collocato describitur, attamen et hoc in situ pylorus et flexus duodeni superior non accurate cognoscuntur, cum ventriculus ipse sursum magis trahendus sit, quo facto pylorus protractus et ex vero situ dimotus perspicitur, ex quibus facile colligi potest, omnibus numeris absolutam iconem duodeni praestari haud posse.

hard to prepare. However, with the artist placed at the feet, and the upper part of the body, especially the chest, somewhat elevated, most of it can be observed, and its lower transverse course from the kidney to the mesocolon and to the beginning of the *jejunum* is perfectly visible. However, both the curvature at the beginning of the duodenum leading from the pylorus, and the concave portion beneath the side of the liver towards the kidney, all the way to the attached descending portion, are not demonstrated well enough. For this reason, another drawing was made, which was sketched with the artist placed on the right side. However, both the position of the pylorus and the superior flexure of the duodenum are not accurately recognized since the stomach itself is pulled more upwards and because the pylorus is seen pulled away and moved from its true position. From this one can easily deduce that, all considered, it is hard to provide a complete picture of the duodenum.

§. V.

"Duodenum a principio suo ad renem descendens describitur"

Duodeni intestini attentam considerationem suscipientibus flexuosus eius tractus, praecipue descensus et ascensus, ita describendus erit, uti ipsius directionem singularem, reliquis visceribus adhuc in situ relictis, deprehendimus. De icone vero ipsis et perfectiore quorundam flexuum definitione, postea sigillatim exposituri sumus, ne filum tractionis nostrae controversiis motis interrumpatur. Ventriculus, eiusque parva extremitas dextrorsum et deorsum directa, in fine suo ita angustatur, ut ruga interne eminens, quam pylori valvulam vulgo dicunt, et quae externe circulo quoque strictiore se prodit (m), eius limites definiat. Immediata tunicarum continuatione, per pylori rugam vix interrupta, canalis noster progreditur, ita tamen, ut ab orificio angustiori in progressu non nihil amplietur, et postquam paululum ascendit, reflectatur et descendat. Uti vero primo hoc tractu retrorsum magis, quam sursum vergere videtur duodenum, sic porro extrorsum potius ad vesiculam felleam usque progreditur, eiusque cervicem et partem corporis tegit, a qua sub hepate recta quasi via defertur, et ad interiorem renis dextri marginem, scilicet ad incisuram usque, quam hylum renalem dicimus, deducitur. In toto hoc tractu, inprimis, si pinguedo circumfusa copiosior est, ipsum duodenum vix conspicitur, sed tantum dextrorsum ad renem, acre paululum ad hunc locum adacto, ut protuberantia cognoscitur, similique modo in pinguedine ulterius pergit.

CHAPTER FIVE

"Description of the duodenum from its beginning to its descent towards the kidney"

The winding course of the duodenum needs to be given careful attention, especially the descending and ascending portions. Thus, this will be described with the rest of the organs left in place so that we can recognize its particular course. Indeed, we will later explain individually the illustration and, in particular, the improved definition of its curves so as to avoid interrupting the thread of our discussion. The stomach, and its small extremity directed to the right and backwards, which is commonly called the *pylorus* valve, is so narrow that its internal folds are prominent while externally it produces a tight ring (**m**) that defines its boundary. The immediate continuation of its layers, with the rugae hardly interrupted along the pylorus, proceeds into the duodenal canal. Yet, after this narrow orifice it becomes somewhat wider in its progression and, after it ascends a little, it turns around and then descends. As the first segment of the duodenum proceeds somewhat posteriorly, it similarly seems to turn upwards, and then continues outwards mainly towards the gallbladder, of which it covers the neck and part of its body. From here it then proceeds downwards under the liver in an almost straight line and extends towards the inner margin of the right kidney, that is, to the notch which we call the *renal hilum*. If the adipose tissue surrounding this tract is abundant, then the duodenum is seen with difficulty, especially towards the right near the kidney where it is reached by a small promontory, known as the protuberance {papilla of Vater}, which similarly travels through adipose tissue.

This chapter included the following footnote:

>>>>>>>>>>>>>>>>>>>>>>>>>>><<<<<<<<<<<<<<<<<<<<<<<<

m) WINSLOVIUS loc. cit. §. 71.

m) see Winslow, cited work, paragraph 71.

>>>>>>>>>>>>>>>>>>>>>>>>>><<<<<<<<<<<<<<<<<<<<<<<<

In 1732, Jacques-Benigne **Winslow** wrote *Exposition Anatomique de la Structure du Corps Humain*. On page 315, paragraph 71, of the "Treatise of the Lower Abdomen," he wrote about the *pylorus* where he commented on its sphincter-like action:

> "La Figure du Pylore est comme celle d'un Anneau transversalement applati, dont le bord interne qui est du côté du Centre, est un peu enfoncé & s'avance dans le Canal Intestinal large & tronqué. Il est naturellement plus ou moins plissé vers ce bord interne, à peu près comme l'ouverture d'une bourse presque serrée. Tout ceci est fort different de ce que les Figures ordinaires & les préparations seches representent. C'est une espece de Sphincter, qui par son action peut retrecir l'Orifice inferieur de l'Estomac, mais ne paroit pas pouvoir le fermer entierement."

> [The shape of the *Pylorus* is like that of a ring, transversely flattened, where the inner edge which is next to the umbilicus is a little sunken and it proceeds wide and shortened into the intestinal canal. It is naturally more or less plicated along its inner surface, a bit like the opening of an almost sealed bag. All this is very different than what is represented in the usual illustrations and dry specimen preparations. It is a sort of sphincter that by its action can narrow the inferior orifice of the stomach, but it does not seem capable of closing it entirely.]

§. VI.

"Ascensus duodeni a rene ad principium ieiuni indicatur"

A commemorata igitur renis parte intestinum nostrum introrsum et sursum vergit, et ita oblique super spinam dorsi, vel potius super venam cavam et arteriam aortam, ut vasa maiora vertebris incumbentia, in sinistram cadaveris partem ducitur. Hoc in tractu pinguis mesenterii radix supra duodenum extenditur, illudque quasi abscondit, ut, nisi extensione quadam facta, observari nequeat, a parte tamen sinistra huius radicis mesenterii duodenum magis in conspectum prodit, quam a dextra, cum in progressu suo dextrum hoc latus relinquat. Dem autem hoc suo tractu ad ultimam dorsi vertebram ascendat, ad mesocolon quoque pertingit, et in duplicatura eius a radice sua pingui suscipitur, a qua tamen, levi flexu facto, tandem in inferiore parte secedit, et antrorsum in ieiunum continuatur. Ultima haec duodeni curvatura sub pancreate haeret, haec enim glandula extremitate sua sinistra et cuspidata a liene aliqua ex parte deorsum extenditur, quae quidem directio in vero situ et illaesis partibus vicinis attendenda, et alia de mesocoli situ repetenda sunt. Cum enim mesocoli ampla membrana fornicem quasi efficiat, diaphragmati non nihil analogum, ideoque abdomen in duas cameras dividat, anterius quidem, inprimis ventriculo alimentis repleto, colon mesocoli situm definiens paululum deprimitur, posterius vero, et ubi duodenum emergit, ad ultimam dorsi vertebram situm est. Mesocolon autem hac sua sede a dextris sinistrosum, hoc est, a concava hepatis parte sursum magis fertur, et ita obliquum situm habet, in eoque loco, ubi duodenum in ieiunum procedit, interdum peculiarem aliquam cavitatem seu partem depressam sistit.

CHAPTER SIX

"The ascent of the duodenum from the kidney to the beginning of the jejunum"

Thereafter, the duodenum proceeds partly inward and upward from the aforementioned kidney, and then obliquely over the dorsal spine, or rather over the vena cava and the aorta since these major vessels actually lie directly over the vertebrae. The duodenum then passes to the left side of the body. Along this route, the fatty root of the mesentery courses over the duodenum and almost hides it, such that, had a certain distention not been made, it would not be observed. However, as it leaves the right side the duodenum comes more into view on the left side of the mesenteric root. While it ascends to the last thoracic vertebra, it also extends toward the mesocolon, and is received into the fold of its fatty root, from which, after making a slight curve, it finally exits in the lower part, and continues forward into the jejunum. This last curvature of the duodenum hugs the bottom of the pancreas, whose pointed left extremity extends towards the dorsal aspect of the spleen, and even its direction and position need to be considered with the nearby organs unaltered. We will now further discuss the position of the mesocolon. Because the wide membrane of the mesocolon acts like a vault, somewhat analogous to the diaphragm, it thus divides the abdomen into two chambers, one anterior, especially with the stomach filled with food and the transverse colon pushed down a little, while the posterior one is where the duodenum emerges at the level of the last thoracic vertebra. The mesocolon however is directed from the right to the left, that is, from the concave aspect of the liver it goes a little upwards, and thus occupies an oblique position and, at the site where the duodenum continues with the jejunum, it sometimes ends in a peculiar cavity or recess {i.e., duodeno-jejunal fossa}.

§. VII.

"Partes, quae duodeno incumbunt, recensentur"

Haec, quae de descensu et ascensu duodeni exposuimus, examine illaesis visceribus vicinis instituto, detecta fuerunt: antequam itaque ad reliqua explicanda progredimur, pauca adhuc de iis, quae intestino nostro incumbunt partibus, disserenda sunt. A pyloro quidem reflexum duodenum ab aliqua et exigua parte lobi sinistri sive minoris hepatis tegitur, et hoc viscere sursum reclinato, principium duodeni optime conspicitur, mox vero membranosae productiones peritonaei, et in obesis potissimum copiosa pinguedo, intestinum hoc nostris oculis subducunt, ita, ut ab extremitate parva ventriculi dextrorsum paululum deducatur, et ibidem lobum dextrum hepatis cum vesicula fellea a latere habeat, de qua vicinia paulo infra nonnulla monebimus. Praecipue tamen ea mesocoli productio, quae a rene sub hepate oblique sursum ascendit, et quae copiosa pinguedine fulcitur, descensum duodeni prorsus obtegit. Coli vero pars ascendens, membranae huic pingui annexa, duodenum non premit, si enim flatu distenta fuerit, antrorsum magis vergit. Notavimus § VI duodenum in ascensu suo super cavam et aortam produci, vasa autem maiora mesaraica in duplicatura mesocoli et ad radicem mesenterii posita, parti huius intestini transversali inferiori incumbunt. Arteria enim ab ortu superior dicta, sub pancreate deinde emergens, mox supra partem duodeni ascendentem ducitur, et in ramos extenditur, qui a trunco sensim in ramos varios mutato, vel potius trunco arcuato, ramos varios exhibente, secedunt, nec a duodeno distento comprimi possunt. Vena quoque mesaraica sub extremitate duodenali panceatis procedens, et postea cum arteria decurrens, a proportione alterius vicinae minoris maior dicta, simili modo in duplicaturam mesenterii producitur, et varie distribuitur. Intestinorum tenuium convolutio in regione umbilicali mesenterio laxe adhaerens, duodeno non incumbit, sed in abdomine fluctuat: glandulae quoque mesaraicae conglobatae ad radicem mesenterii dispersae, dummodo scirrhosae non sunt, intestinum arcte licet comprehensum offendere nequeunt.

CHAPTER SEVEN

"A list of organs adjacent to the duodenum"

The organs adjacent to the duodenum, which we described in the chapters on the ascent and descent of the duodenum, were revealed in an investigation undertaken with intact neighboring organs. Before we continue, however, a few of these need to be discussed. Indeed, the duodenum, as it turns away from the pylorus, is protected by a small portion of the left lobe of the liver, and the beginning of the duodenum is best seen with the liver lifted upwards, but then the membranous layers of the peritoneum — which contain abundant adipose tissue in the obese — obscure this intestine from our eyes. Then, as the duodenum extends from the small end of the stomach a little to the right, it has to its side the right lobe of the liver and the gallbladder, the vicinity of which we will deal with a little later. Notably, however, the layer of the mesocolon, which arises obliquely from the kidney under the liver, is strengthened by a great quantity of adipose tissue and completely covers the descent of the duodenum. Indeed, the ascending colon, connected to this fatty membrane, does not press against the duodenum unless distended by gas, and tends more towards the front. We noted in chapter 6 that the duodenum in its ascent proceeds across the aorta and vena cava, but the greater mesenteric vessels situated in the double layer of the mesocolon and the root of the mesentery lie over the lower transverse part of this small intestine. Indeed, the superior mesenteric artery emerges from under the pancreas, courses over the ascending portion of the duodenum, and is divided into branches, as it gradually changes from a trunk into numerous branches {intestinal arteries}, or rather an arched trunk {middle colic artery} exhibiting numerous branches which spread out so that they cannot be compressed by a distended duodenum. And the {superior} mesenteric vein proceeding under the duodenal end of the pancreas, and after running with the artery, called "greater" in relation to the nearby smaller ones, proceeds in a similar manner into the fold of the mesentery, and is variously distributed. The convolutions of the small bowel which are loosely attached to the mesentery in the umbilical region do not lie upon the duodenum but float about in the abdomen. And the conglobate mesenteric lymph nodes, which are dispersed along the root of the mesentery, are unable to harm the tightly surrounded intestine, provided that they are not hard.

§. VIII.

"Pinguedo duodenum cingens consideratur"

Pinguedo, quae passim abdominis visceribus intersternitur, et in maioribus vel minoribus telae cellularis interstitiis colligitur, circa duodeni quoque tractum copiosa conspicitur, et quanquam in macilentis et iunioribus, ut ubique, sic et hic parcior illa deprehendatur, et magis cellulares tractus, et membranae ligamentosae, quam vera pinguedo detegantur, ideoque demonstratio duodeni melior fiat, quam in obesis, tamen cum in macilentis substantia adiposa in abdominis non prorsus deficiat, duodeni decursum accurate satis persequi non possumus, nisi simul pinguedinis mentionem iniiciamus. In obesis omentum parvum, quod minorem ventriculi curvaturam succingit, ad hepar producitur, inde supra duodenum in descensu usque ad colon pertingit, copiosaque pinguedine duodenum non tantum ab anteriore parte comitatur et involvit, sed etiam alia productione membranae cellularis in posteriori et interna parte eidem substernitur (n). Non emergit hic duodenum, sed in cellulari et pingui mesocoli substantia ulterius usque ad flexum inferiorem progreditur, et cum pinguedine renum et radice pingui mesocoli supra renem ascendentis ferme coalescit. Ab hoc vero loco, dum iterum sub pingui mesenterii membrana ad radicem mesocoli transversi, et initium ieiuni sursum ducitur, quasi in vagina delitescit. Quamvis enim et hic, ut in toto ferme tractu, duodenum membranam externam, a peritoneo recipiendam, pressius adhaerentem non habeat, ambitus tamen membranae cingentis laxior ita comparatus est, ac si per vaginam quandam ex mesocolo et principio mesenterii formatam produceretur (o).

CHAPTER EIGHT

"A consideration on the fat that surrounds the duodenum"

Adipose tissue, which is distributed everywhere among the abdominal organs and is gathered in the greater or smaller cellular interstices of the tela {subserosa}, is also seen abundant along the course of the duodenum. However, in the gaunt and the young, this is found more sparingly like everywhere else, both in the more cellular tract and the ligamentous membranes, which indeed are not covered by fat, and thus improves the demonstration of the duodenum. In the obese instead, in whom the adipose tissue in the abdomen is not altogether lacking as in the gaunt, we cannot follow the course of the duodenum accurately enough, unless we first mention about the adipose tissue. In the obese, the *lesser omentum,* which encircles the lesser curvature of the stomach, reaches the liver, and from there extends over the duodenum all the way down to the colon. The duodenum is accompanied and wrapped around by abundant adipose tissue, not only along its anterior side but also by the production of cellular membranes spread beneath the posterior and inner side of the same (**n**). The *duodenum* does not emerge here but proceeds into the substance of the cellular and fatty mesocolon and further towards the lower flexure, and almost coalesces with the fat of the kidney and with the ascending root of the fatty mesocolon rising above the kidney. Indeed, from this place it yet again hides as if in a sheath under the membrane of the fatty mesentery at the root of the transverse mesocolon, and travels upward to the beginning of the jejunum. For even here, as in almost its entire course, the duodenum does not have a closely adherent external membrane derived from the peritoneum and the space created by the surrounding membranes is arranged looser, as if a sort of sheath were formed by the mesocolon and the beginning of the mesentery (**o**).

74

This chapter included two footnotes, "n" & "o":

>>>>>>>>>>>>>>>>>>>>>>>>>><<<<<<<<<<<<<<<<<<<<<<<

n) Asserto MONROI hic magis insistimus, cum de obesis sermo sit.
    HALLERUS quidem Tab. II. omenti vid. iconum fasciculum I.
    Gottingae 1743. fol. lit. Y tantum lubricam et flavam
    membranam adducit, quae, ut ligamentum a vesicula fellea
    supra duodenum, cui externam membranam largitur, ad colon
    ducitur, qualis in macilentis observatur.

n) We apply here the assertion of **Monro**, in his discussion about
    the obese. **Haller**, in fact, in Table 2 of the omentum – see
    Iconum, fasciculus I, Gottingae (1743), folio, letter "Y" – shows
    a very greasy and yellow membrane which, like a ligament,
    extends from the gallbladder over the duodenum, to which it
    bestows an external membrane, and reaches the colon, as one
    observes in the thin individual.

>>>>>>>>>>>>>>>>>>>>>>>>>><<<<<<<<<<<<<<<<<<<<<<<

This footnote refers to the 1st fascicle of *Iconum Anatomicarum
Partium Corporis Humani* written in 1743 by **Haller**. His Figure #2
was dedicated to the omentum and a closer view reveals the area
specified by Claussen (see figure on next page):

In Haller's figure, the letter Y corresponds to:

Y. Ligamentum sive Membranae e), quae a vesicula &
continuo sulco transverso ad colon eunt, trans Duodenum,
cui sunt pro externa membrana, & aliqua parte adhaerent.

[Y. Ligament or membranes e), which proceed from the
gallbladder as a continuous transverse sulcus to the colon,
across the duodenum, to which they are like an external
membrane, and adhere to other organs.]

Haller inserted a pertinent footnote here ("e") which said:

e) Has WINSLOWUS ab hepate & vesicula oriri, sed in
duodeno terminati ait. IV, tr. du bas ventre n. 359. Mem. de
l'Acad. 1715. p. 317. MONROUS vocat duplicaturam omenti

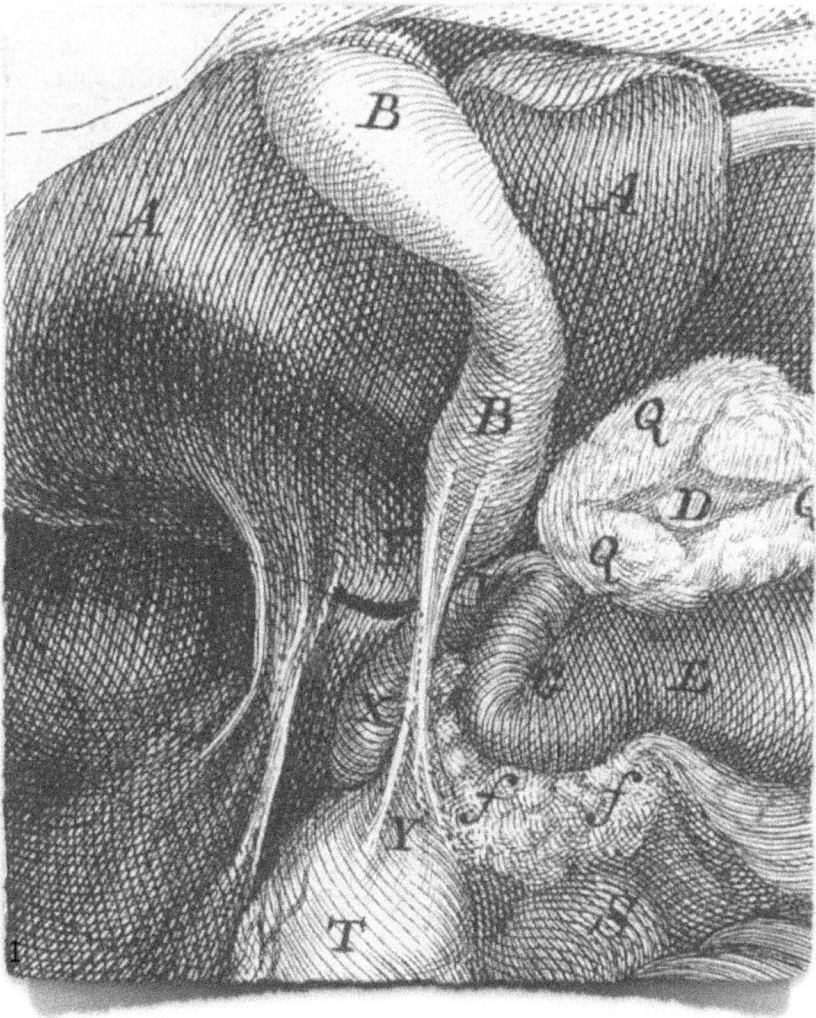

Iconum Anatomicarum Partium Corporis Humani written by Haller (1743).

Other noteworthy lettered organs are:

A = liver; B = gallbladder; D = lobe of Spigelius; E = stomach; G = pylorus;
Q = lesser omentum; S = mesocolon; T = colon; V = 2nd flexure of the duodenum;
X = 3rd flexure or descending portion of the duodenum; f = pancreas.

l. c. G.H. Verum simplex ibi est, & flava, et lubrica membrana, continua capsulae GLISSONII et omento minori.

[**Winslow** says that these arise from the liver and gallbladder but terminate in the duodenum. *Traite du bas* *ventre*, paragraph 359; *Memoire de l'Academie Royale*, page 317 (1715). **Monro** called it a fold of the omentum; cited work, letters G and H. In truth, there is a single membrane, both yellow and greasy, which is continuous with Glisson's capsule and the lesser omentum.]

Haller's footnote adds two additional references by Wilson, one from 1732 and the other from 1715.

In 1732, Jacques-Benigne **Winslow** wrote *Exposition Anatomique de la Structure du Corps Humain*. On page 544, paragraph 359, of the "Treatise of the Lower Abdomen," he wrote about the *epiploon*:

"**Ainsi au moyen de l'intervalle de l'Estomac & du Mesocolon les deux Epiploons ne font ensemble qu'une seule capacité commune, laquelle s'ouvre dans la cavité du Bas-Ventre par un seul orifice commun, situé près de la Commissure du côté droit du grand Epiploon. Cet orifice est semilunaire ou demi-circulaire, & formé par l'union des deux Ligamens Membraneux, dont l'un attache au Foye le commencement du Duodenum & le Col de la Vesicule Biliare, l'autre y attache la portion voisine du Colon & s'étend jusqu'au Pancreas. Il en résulte un bord en maniere d'Anse qui embrasse la Racine du Lobule, en laissant autour de cette Racine une ouverture assez large pour y passer le bout d'un doigt.**"

[Thus, midway between the stomach and the mesocolon the two omenta together make just one common cavity, which opens into the cavity of the lower abdomen through a single common opening, located near the commissure of the right edge of the greater omentum. This opening is semilunar, or semicircular, and is formed by the union of two membranous ligaments, one which connects the beginning of the duodenum and the gallbladder neck to the liver; the other is attached to the adjacent portion of colon and extends to the pancreas. The result is an edge like a loop that embraces the base of the caudate lobe {of Spigelius}, leaving around this base an opening that is large enough to allow the passage of a finger tip.]

In 1715, **Winslow** wrote the article "Nouvelles Observations Anatomiques sur la Situation et la Conformation de Plusieurs Visceres" in the journal *Mémoires de l'Academie Royale*. On page 234 (not 317), Winslow wrote about the *epiploon*:

"Je fus assez heureux de trouver une ouverture naturelle très-considérable, & en même tems fort surpris de la voir située dans un endroit, sur lequel on passe très-souvent sans y faire attention; sçavoir sous le le grand lobe du Foie entre un ligament membraneux qui lie le commencement du Duodenum conjointement avec le col de la vésicule du fiel au Foye, à côté d'une éminence qui est comme la racine du petit lobe de Spigelius, & un autre qui attache le Colon avec le Pancréas."

[I was fortunate enough to find a very considerable natural opening, and at the same time was greatly surprised to see it located in a place which one very often pays little attention to, namely, under the right lobe of the liver within a membranous ligament that connects the beginning of the duodenum as well as the neck of the gallbladder with the liver, next to a hillock which is like the base of the small lobe of Spigelius, and another that attaches the colon to the pancreas.]

Haller, Winslow, Monro, and consequently also Claussen, are referring to the so-called ligament (Y) that goes from the gallbladder (B) to the hepatic flexure of the colon (T) passing over the duodenum (X). Today, this would correspond to the *hepatoduodenal* or the *cystoduodenal ligament* and the *duodenocolic ligament*, or even the *cholecysto-duodeneno-colic ligament*. The hepatoduodenal ligament contributes to the right free margin of the lesser omentum and thus the anterior boundary of the epiploic foramen (of Winslow). A special prolongation of the hepatoduodenal sometimes extends downward to the transverse colon, forming the *hepatocolic ligament*. These so-called ligaments are actually just folds or duplications of the peritoneal membranes, and the hepatoduodenal segment notably carries the common bile duct.

78

>>>>>>>>>>>>>>>>>>>>>>>>>><<<<<<<<<<<<<<<<<<<<<<<<

o) Hanc vaginae singularis speciem triangulare spatium cellulari tela repletum thecae (etui) ad instar dispositum WINSLOVIUS dicit l. c. §. 106 et 107. alibi vero triangularem canalem (tuyau) duodeni appellat l. c. §. 204 MONROUS l. c. pag. 68 annulum dicit, ex quo duodenum procedit.

o) This triangular space is a unique form of sheath filled with a layer of cellular substance which **Winslow** called a case (*etui* in French) from the manner of its arrangement – cited work, paragraph 106 & 107; elsewhere he named it the triangular pipe (*tuyau* in French) of the duodenum; cited work, paragraph 204. **Monro** called it a ring from which the duodenum proceeds; cited work, page 68.

>>>>>>>>>>>>>>>>>>>>>>>>>><<<<<<<<<<<<<<<<<<<<<<<<

In 1732, Jacques-Benigne **Winslow** wrote *Exposition Anatomique de la Structure du Corps Humain*. See page 41 for the transcription of the passages from § 106 and 107 of Winslow's "Treatise of the Lower Abdomen." As for § 204, Winslow discusses the anatomy of the mesocolon and says:

> "Dans tout ce trajet le Mesocolon s'élargit, & forme un Plan demi-circulaire presque transversal, & très-peu plissé vers la circonference du grand bord. Il est attaché par ce grand bord tout le long de l'Arc du Colon, & par là cache une des Bandes ligamenteuses de cet Intestin, sçavoir celle de la petite convexité de l'Arc. Il forme par le petit bord le Tuyau triangulaire du Duodenum, & produit par le grand bord la Tunique externe du Colon, de la même maniere que le Mesentere fait celle des Intestins grêles. En passant sous la grosse extrémité de l'Estomac, il est un peu adhérant à la portion inferieure de cette extrémité, qui par sa portion superieure l'est aussi au Diaphragme."

[Throughout its course the mesocolon widens, and becomes an almost transverse semicircular surface, and is minimally folded towards the circumference of its wide edge. It is attached along this wide edge by the entire length of the arc of the {transverse} colon, and thus hides one of the ligamentous bands of this intestine, namely that of the small convexity of the arc. Along the narrow edge it creates the triangular pipe of the duodenum, while along the wide edge it forms the outer {serosal} layer of colon, in the same manner as the mesentery does for the small intestines. While passing under the greater curvature of the stomach, it is somewhat adherent to the inferior portion of this, while the superior aspect {of the stomach} adheres to the diaphragm.]

As for the **Monro** reference, this is from his article "The Description and Uses of the Intestinum Duodenum" in volume 4 of the 2nd edition of *Medical Essays and Observations*. We have reprinted this in pages 44-48 above, and will here repeat the relevant passage:

"After the Curve just now described, the Duodenum is involved in the Root of the Mesentery, and mounts obliquely within it towards the left Side, with the Vena Cava behind it; and after a Course of about four Inches there, rises forwards, to acquire a proper Mesentery, or to commence Jejunum, the Membranes of the Root of the Mesentery seeming to make a Ring at which the Gut comes out, though they are really continued on the Intestine, and forms its external membranous Coat."

§. IX.

"Limites duodeni desiniuntur"

De limitibus duodeni dubium quoque oritur, cum ipsa mensura duodecim pollicum a veteribus assumta in singulari huius intestini flexu difficile nec satis certo applicetur, et praeterea intestinorum quod longitudinem dimensio pro differentia subiectorum omnino variet. Qui post Vesalium (p) recta descendere duodenum assumserunt, illi ideo limites eius ad insertionem ductus biliaris communis posuerunt. Hinc etiam Boerhaavius (q) ductum biliarium et pancreaticum in duodenum ad finem rectissimi sui decursus inseri, praecedentium anatomicorum auctoritate seductus, asservit. Recentiores vero anatomici, qui in situm huius intestini accuratius inquisiverent, et a nobis nominati fuerunt § II in eo consentiunt omnes, limites duodeni in eo loco optime figi posse et debere, ubi, post varios flexus absolutos, tandem ex mesocoli duplicatura in inferiorem abdominis cameram producitur, et ubi ex mesocolo emergens mesenterii membrana, intestinum tenue suscipiens, aliquo modo exsurgit, et ad ulteriorem ieiuni progressum firmandum producitur, licet enim Glissonio (r) de lana caprina certare videantur dissectores, dum hi ingressum ductus cholodochi ad duodenum finem, alii autem ad ieiuni principium statuant, ex ipso tamen situ et nexu, quem declaravimus, limites satis certi et naturae convenientes figuntur, qui etiam ex usu, inferius exponendo, patebunt. Duodenum autem est prima illa intestinorum tenuium pars, quae ob flexum suum alimenta semidigesta ad tempus retinet, lentius movet et digerit, celerique postea progressu in ieiuni convolutiones propellit. Negandum quidem non est ad insertionem ductus cholodochi et pancreatici in hominibus semper fere concurrentis, illos humores, quibus digestio ulterior in duodeno praestanda, perficitur, accedere, sed causa remorae, sub qua illa mutatio fieri potest, est singularis intestini progressus, et continuatio ascensus usque ad ultimam dorsi vertebram.

CHAPTER NINE

"Designation of the boundaries of the duodenum"

The question also arises regarding the ends of the *duodenum* because its length of twelve inches obtained from the Ancients is difficult to apply with certainty to this particular curved intestine and, furthermore, the length of the intestines varies among different subjects. Those who assume, just as **Vesalius** did (**p**), that the *duodenum* descends in a straight line, consequently place its ending at the insertion of the common bile duct. Hence, also **Boerhaave** (**q**), seduced by the authority of previous Anatomists, claimed that the biliary and pancreatic ducts are inserted into the duodenum at the end of its very straight course. However, more recent anatomists (and these have been named by us in chapter 2), who made more accurate observations as to the location of this intestine, all agree that the end of the duodenum can and should be well determined in that place where, after several complete curvatures, a chamber is ultimately created from the duplication of the mesocolon in the lower abdomen. As the small bowel emerges from the mesocolon, the mesenteric membane rises up a little and produces a reinforcement to the further progression of the *jejunum*. **Glisson** even says (**r**) that dissectors certainly seem to argue about trifles as long as some argue that the entrance of the bile duct is at the end of the duodenum while others place it at the beginning of the jejunum. Nevertheless, from its very position and relation, which we have indicated, the boundaries are quite certainly fixed, the lower of which will also be revealed from experience as will be explained below. The *duodenum* is the first part of the small intestines, which retains the semidigested food because of its flexures, and slowly moves and digests it, and propels it rapidly into the loops of the *jejunum*. It is not to be denied that the biliary and pancreatic ducts in humans almost always meet at the insertion, and their fluids, which further aid the digestion in the duodenum, complete it. But the cause of delay under which that change can be performed is due to the unusual path of this intestine and the continuation of the ascent as far as the last thoracic vertebra.

This chapter included the following three footnotes "p", "q", and "r":

>>>>>>>>>>>>>>>>>>>>>>>>>>><<<<<<<<<<<<<<<<<<<<<<<<

p) loc. cit. p. 425.

p) cited work, page 425.

>>>>>>>>>>>>>>>>>>>>>>>>>>><<<<<<<<<<<<<<<<<<<<<<<<

In 1725, the work by Andreas **Vesalius** *De Fabrica Corporis Humani*, originally written in 1543, was re-issued in the collected work called *Opera Omnia Anatomica & Chirurgica* by Boerhaave and Albini. On page 425 of volume 1, Vesalius talked about the *duodenum* as the first part of the small intestine and specified that it had a straight course:

> **"Primum quidem constituitur tota ea intestini pars, quae ventriculo substrata, ab inferiore ipsius orificio recta quodammodo eousque descendit, ubi intestinum in anfractus orbesque primum convolvi incipit."**
>
> [In fact, the first intestine is made up of the entire segment of intestine which, positioned beneath the stomach, in a certain manner descends straight down from its lower opening all the way to where the intestine first starts to roll up into coils and loops.]

>>>>>>>>>>>>>>>>>>>>>>>>>>><<<<<<<<<<<<<<<<<<<<<<<<

q) vid. Instit. medic. §. 96 et comment. HALLERI Vol. 1. pag. 408. qui sententiam BOERHAAVII non tuetur.

q) see *Institutiones Medicae*, chapter 96, with comments by Haller, volume 1, page 408, who does not support the statement of Boerhaave.

>>>>>>>>>>>>>>>>>>>>>>>>>>><<<<<<<<<<<<<<<<<<<<<<<<

In 1740, Albrecht **Haller** edited and annotated volume 1 of the *Praelectiones Academicae in proprias Institutiones Rei Medicae* written by Herman **Boerhaave**. His comments on page 408 of chapter 96 have been previously addressed (see page 49-51).

>>>>>>>>>>>>>>>>>>>>>>>>>><<<<<<<<<<<<<<<<<<<<<<<<

r) vid. Anatomia hepatis. Edit. Amstelodami 1665. in 12. cap. XVI. p. 135.

r) see *Anatomia Hepatis*, Amsterdam edition (1665) in 12mo, chapter 16, page 135.

>>>>>>>>>>>>>>>>>>>>>>>>>><<<<<<<<<<<<<<<<<<<<<<<<

In 1665, Francis **Glisson** wrote *Anatomia Hepatis*. On page 135, he wrote:

"Ductus communis a praedictis vasis exorsus, recta deorsum tendit versus finem duodeni, aut (si mavis) jejuni initium; nam eodem haec res redit; est enim utriusque intestini limes sive terminus. Et licet intestina haec aliter quoque ab invicem distinguantur (nempe, quod duodenum nullis spiris circumplecitur, utpote mesenterio minime annexum) ab hujus tamen vasis ingressu facillime internoscuntur. Ideoque mihi de lana caprina certare videntur, dum hi ingressum ejus ad duodeni finem, illi autem ad jejuni principium statuunt."

[The common bile duct that began from the aforementioned ducts courses straight downwards toward the end of the *duodenum*, or (if you prefer) the beginning of the *jejunum*; for this same matter comes up again; it is truly the boundary or limit of both intestines. And although these intestines can also be distinguished from each other (namely, that the duodenum is not folded around in coils, since it is not attached to the mesentery), nevertheless, they are easily distinguished by the entrance of this duct. Therefore, it seems to me that the {Anatomists} argue about trifles, as long as some of them place its entrance at the end of the duodenum and others instead at the beginning of the jejunum.]

§. X.

"Situs vesiculae felleae iuxta duodenum attenditur"

His praemissis icones a nobis exhibitas accuratius explicare, et cum aliis a citatis auctoribus, § II expositis, conferre e re nostra esse arbitramur. Summariam quidem partium in iconibus delineatarum recensionem in fine subnexuri erimus, quo illae uno intuitu perspici queant, in hac vero tractatione ad partes duodeno vicinas praecipue nostram attentionem convertimus, quo huius intestini nexus et situs eo melius patescat. Vesicula nimirum fellea a latere dextro duodeni, lobo scilicet magno hepatis, affixa in nostris iconibus tantum obscure indicatur: in priore enim ad lit. e) vix apparet, in altera autem, ubi ad lit. b) plenius conspici deberet, flaccida propendet. Quicunque autem verum hepatis situm perpendit, et simul cognoscit flaccidam fuisse vesiculam, is facile concedet, eam ita depictam esse, uti oculis delineatoris sese obtulit: in icone vero Garengeotti ad lit. F in loco nimis lucido comparet, in Monroi autem delineatione totum hepar ad lit. AA ob costas spurias, nimis distentas, ex situ genuino deductum et vesicula fellea ad lit. C perverso et perpendiculari situ disposita deprehenditur. Tenendum vero est, nos vesiculam hanc, flatu extensam et elevatam, ostendere noluisse, quoniam sine laesione cervicis, sub duodeno latentis, illud perficere haud potuissemus, in icone tamen adornanda intestinum nostrum ex situ suo dimovere inconsultum erat, vesicula itaque fellea post flexum duodeni superiorem eidem accumbit, illudque colore luteo transparente saepius tingit, ast magis producta et ad fundum procedens pars, coli zonae ascendenti, quae in icone prima remota est, incumbit, et interdum, si limbum hepatis transgreditur, vix vero in adultis superat, nisi flatu distendatur, plus eminet, quod, cum in cadavere non conspectum fuerit, indicari quoque haud potuit.

CHAPTER TEN

"Position of the Gallbladder near the Duodenum"

We believe that the aforementioned illustrations presented by us, along with others from all the authors mentioned above, and introduced in chapter 2, can more accurately explain and contribute to our discussion. We have added a list of the organs drawn in the illustrations at the end of the review, so that we can see at a glance where they are, but in this discussion, we will turn our attention especially to the neighboring organs of the duodenum, and reveal their relation and position to this intestine. The gallbladder on the right side of the duodenum is indicated rather indistinctly attached to the large lobe of the liver in our drawings: in the first figure (as the letter "e") it is hardly visible, but in the second, when it should be more noticeable (as letter "b"), it hangs down flaccidly. Yet, whoever examines the true position of the liver, and at the same time recognizes that the gallbladder is flaccid, allows it to be easily depicted in this way as it presents itself to the eyes of the artist. It is clearly visible in the picture of **Garengeot** at the letter "C." However, in the drawing of **Monro** the entire liver at letter "AA" is observed at the level of the false ribs, which are pulled outwards, and removed from their genuine position; and the gallbladder at letter "C" is found to be erroneous and arranged in a perpendicular position. Bear in mind, however, that we could not demonstrate this gallbladder lifted up and expanded by gas without any damage to its neck, which lies beneath the duodenum. It was nevertheless ill-advised to remove the duodenum from its position while preparing the drawing, and thus the gallbladder, which often develops a transparent yellow color, is lying against the duodenum after its superior flexure, but its dilated portion reaches both downwards and overhangs the zone of the ascending colon, which is removed in the first drawing. And sometimes, if the gallbladder extends beyond the edge of the liver, which in fact it barely passes in adults unless it is distended by gas, it protrudes much more, but could not be shown since it was not observed in the cadaver.

§. XI.

"Pancreas, quatenus duodeno adhaeret, consideratur"

Simili quoque modo a sinistra et interiore parte duodeni descendentis posita pancreatis glandula considerari meretur, quippe quae in spatio angusto, quod intra inferiorem flexum formatur, extremitate sua dextra delitescit, nuda enim non apparet, sed a pingui vel cellulosa tela radicis mesocoli contegitur; ideoque necesse est, hanc mesocoli partem paululum removere: sed et hac portione remota, extremitas pancreatis dextra sive duodenalis tota non conspicitur. Parvum enim illud (s), quod GRAAFIUS iam in icone sua eleganter depinxit, et in quo peculiarem et reliquis ductibus lateralibus maiorem indicavit canalem (t), in abscondito potius et inferiore loco haeret. Obiter hic monemus parvum hoc pancreas ut peculiare et a magno distinctum viscus vix assumendum esse, quoniam cum magno firmiter cohaeret, et continuata acinorum serie, uti reliquae pancreatis partes, coniungitur. MONROUS quidem in icone sua et magnum lit. M et parvum pancreas lit. N, cum ductibus suis ad duodenum progredientibus depingere annisus fuit, mesaraicamque arteriam et venam per intercapedinem utriusque pancreatis descendentes lit. O, nec non ductum cholodochum communem lit. P oculis sistere allaboravit, sed praeter id, quod delineationi magna obscuritas accedat, facile cognoscitur, has partes omnes in conspectum non deduci posse, nisi pancreatis substantia dissecta et duodeno non nihil e sede remoto, quod autem inferiorem hunc flexum naturae convenientem non sistit; praeter extremitatem vero duodenalem nihil conspici potest, cum haec glandula, praecipue quoad mediam sui partem, triquetra, mox sub ventriculum se abscondat, et ita prorsus oblique sursum feratur, uti illam in nexu cum aliis visceribus consideratam, § VI iam aliqua ex parte descripsimus.

CHAPTER ELEVEN

"Considerations on the Pancreas, and how much it adheres to the Duodenum"

In a similar manner, the pancreas situated on the left and on the inside of the descending duodenum deserves to be considered, since its right extremity is hidden in the narrow space which is formed within the lower bend. It does not appear naked, but it is covered by the fatty or cellular membrane of the mesenteric root. Therefore, it is necessary to remove a small portion of this mesocolon; but when this is removed, the right or duodenal extremity of the pancreas is not completely visible. In fact, the lesser pancreas (**s**), which **de Graaf** elegantly depicted in his illustration, and in which he indicated the special major duct and remaining lateral ducts (**t**), lies in a rather hidden and deep location. Incidentally, here we remark that the lesser pancreas can hardly be assumed to be unique and distinct from the great organ, because it is firmly adherent to the greater pancreas and their acini are connected as a continuous series, like other parts of the pancreas. In his illustration, **Monro** strived to depict the greater pancreas with the letter "M" and the lesser pancreas with the letter "N," with its ducts proceeding to the duodenum, the mesenteric artery and vein both descending through the interval of the pancreas at letter "O," and he also attempted to show the common bile duct at letter "P." Yet, in spite of the great obscurity of the drawing, it is easy to recognize that all these organs cannot be brought into view unless the substance of the pancreas is dissected away and the duodenum is removed somewhat from its site, which however does not end appropriately and naturally at this lower flexure. In addition, the duodenal end of the pancreas cannot be seen since this triangular-shaped gland, especially its middle part, immediately hides itself under the stomach, and then is directed obliquely upwards, although with regards to its relation with other viscera, we have in some measure already described this in chapter 6.

This section included two footnotes "s" and "t":

>>>>>>>>>>>>>>>>>>>>>>>>>>>><<<<<<<<<<<<<<<<<<<<<<<<

s) Hanc partem pancreas parvum appellat WINSLOWIUS, traité du bas ventre § 324 ab ilia dicitur caput, sed placuit cum HALLERO principium huius glandulae a liene assumere, vid. Physiol. pag. 454. Gott. 1751. 8.

s) Winslow calls this organ the "small pancreas" in his *Traité du Bas Ventre*, paragraph 324, while others call it the "head" {of the pancreas}, but he agrees with Haller in assuming that the beginning of this gland was from the spleen, see *Physiologia*, page 454, Gottingae (1751), octavo.

>>>>>>>>>>>>>>>>>>>>>>>>>>>>><<<<<<<<<<<<<<<<<<<<<<<<

In 1732, Jacques-Benigne **Winslow** wrote the section dedicated to the anatomy of the lower abdominal organs "Traité du Bas Ventre" in his book *Exposition Anatomique de la Structure du Corps Humain*. The chapter #324 was entitled:

"Le Petit Pancreas"

"J'ai trouvé il y a plusieurs années dans l'Homme la grosse extrémité du Pancreas à l'endroit où elle est attachée à la courbure du Duodenum, faire une espece d'allongement en bas collé sur la portion suivante de l'Intestin. En l'examinant j'y ai trouvé un Conduit Pancreatique particulier, ramifié comme le grand Conduit, qui se portoit vers l'extrémité du grand, se croisoit avec lui, & ensuite perçoit le Duodenum & s'ouvroit dans l'extrémité du grand Conduit. J'appelle cette portion le petit Pancreas. Quelquosois il s'ouvre aussi séparément dans le Duodenum, dans lequel on trouve aussi quelquesois plusieurs petits Trous presque imperceptibles autour du Canal Cholidoque, lesquels Trous répondent au Pancreas."

[Several years ago, I found that in humans the large extremity of the pancreas on the right where it is attached to the curvature of the duodenum presents a sort of downward extension attached to the subsequent portion of this intestine. In my investigation I found a peculiar pancreatic duct which was branched like the main pancreatic duct and travelled towards the end of the main pancreatic duct, crossed over it, and then perforated the duodenum and opened at the end of the main pancreatic duct. I named this part the *small pancreas*. Sometimes it opens separately into the duodenum, in which one finds several almost imperceptible small openings around the common bile duct, openings that derive from the pancreas.]

This particular "organ" will eventually be known as the "uncinate process" of the pancreas. It is also known as the "ventral pancreas" which embryologically arises from a different epithelial bud than the larger dorsal pancreas. In fact, it originates on the right side of the enteric tube and then swings posteriorly to the left and eventually merges with the "dorsal pancreas." Its duct will merge with the duct of Wirsung and will empty into the major duodenal papilla or papilla of Vater. The duct of the dorsal pancreas will often merge with that of the "petit pancreas," although occasionally it will persist as the duct of Santorini and empty into the minor duodenal papilla situated proximal to the major papilla of Vater.

>>>>>>>>>>>>>>>>>>>>>>>>><<<<<<<<<<<<<<<<<<<<<<<<<
t) vid. loc. cit. Tab. I. lit. F.

t) see Figure 1, letter F, of the cited work.

>>>>>>>>>>>>>>>>>>>>>>>>><<<<<<<<<<<<<<<<<<<<<<<<<

Claussen refers us to figure 1 in the book *De Succi Pancreatici* written by Reinier de **Graaf** from 1664, which contained an illustration depicting the *duodenum* (see picture on page 37). Letter "F" corresponds to the pancreas.

§. XII.

"Flexus duodeni primus et superior describuntur"

Officio nostro satisfecisse putamus, cum descensum atque ascensum vicinasque duodeni partes ample satis descripserimus, lucem vero huic doctrinae maiorem accedere opinamur, si praecipuos duodeni flexus curatius paulo examinemus. De superiore sive primo flexu dicturi, extremitatem ventriculi parvam, in qua prope pylorum antrum quasi describit WILLISIUS (u), iam quodammodo sursum et retrorsum inclinari certum est, hinc prima ad pylorum observata curvatura exigua tantum est, quae, etsi aer in ventriculo contentus, per orificium pylori angustatum ad principium duodeni urgeatur, ad pollicis altitudinem namquam accedit, dummodo vincula vicina illaesa sint, si autem a reliquis partibus connexis hoc duodeni principium separatur, tunc multum cedit, et ita quidem flatu distendi potest, ut interdum ultra pollicem ascendere videatur. Quoniam itaque pylorus cum principio duodeni versus corpora vertebrarum reflectitur, et sursum progreditur, numquam tamen ad eam, quae cardiae est, altitudinem pertingit, licet parum absit, quin eo deferatur. Ex nostra enim observatione cardiam sive orificium superius ventriculi ad sinistram et paulo anteriorem partem cartilaginis inter decimam et undecimam dorsi vertebram positae (w). et pylorum, sive inferius orificium a latere dextro cartilaginis inter undecimam et duodecimam vertebram parum distare conspeximus: quamvis non negemus, in cadavere aperto et hepate mole sua descendente pylorum ex vero suo situ paululum deorsum dimoveri. Quoniam itaque extremitas parua ventriculi cum hoc ipso flexu duodeni sub concava hepatis parte collocatur, duodenum post flexum descendens, aliqua ex parte transversim ad vesiculam felleam usque progreditur. Hic autem transversus progressus nobis non tantus visus fuit, qui peculariem distinctionem exigerit, sed vere asserere possumus, intestinum, hoc flexu facto, mox descendere, et a vesiculae felleae latere procedens in pinguedine mesocoli ascendentis abscondi et ad renem pergere.

CHAPTER TWELVE

"Description of the First and Superior Curvature of the Duodenum"

We believe that we will fulfill our duty if we fully describe the ascent and descent of the duodenum as well as its neighboring organs. Thus, we decided to shed more light on this by examining a little more in depth the curvatures of the duodenum. Regarding the superior or so-called first curvature, it is certain that the small extremity of the stomach, which near the pylorus was described by Willis as resembling a chamber {*antrum*} (**u**), is bent somewhat upward and backward. Here, one first observes only a slight curvature after the pylorus, which rises to the height of an inch when the air contained in the stomach pushes through the narrow opening of the pylorus to the beginning of the duodenum, provided the relation with nearby organs is not altered. But if the beginning of the duodenum is separated from the other attached organs, it then yields much more and can thus be distended with air, so that sometimes it can be seen rising more than one inch. Therefore, while the pylorus with the beginning of the duodenum turns back towards the vertebral bodies and proceeds upwards, it never reaches the height of the cardia, even though it is carried not too far from it. Indeed, from our observations the *cardia* or upper orifice of the stomach is located a little to the left and in the front of the cartilage between the 10th and the 11th thoracic vertebra (**w**); and the *pylorus* or the lower opening can be seen not far from the right side of the cartilage between the 11th and 12th vertebra, although we would not deny that in the opened corpse the pylorus is shifted a little backwards from its true position by the descending mass of the liver. And thus, since the small end of the stomach with the curvature of the duodenum is located under the concave part of the liver, the duodenum after this curvature descends and proceeds somewhat across the gallbladder. Here, however, we do not see so much a transverse advancement, which is a distinction that needs to be made, but in truth we can affirm that the intestine, having made this curvature, soon afterwards descends and, proceeding from the side of the gallbladder, hides in the fat of the ascending mesocolon, and continues towards the {right} kidney.

....

This chapter contained two footnotes "u" and "w":

>>>>>>>>>>>>>>>>>>>>>>>>>>>><<<<<<<<<<<<<<<<<<<<<<<<<

u) **De medicamentorum operationibus Sect. I. Cap. 2. pag. 8. edit. opp. Amstel. 1682. 4to.**

u) *De Medicamentorum Operationibus,* section 1, chapter 2, page 8, Amsterdam edition (1682), quarto.

>>>>>>>>>>>>>>>>>>>>>>>>>>>><<<<<<<<<<<<<<<<<<<<<<<<<

In 1674, Thomas **Willis** (1621-1675) wrote the book *Pharmaceutice Rationalis sive Diatriba De Medicamentorum Operationibus in Humano Corpore.* It was later republished in 1682 within the anthology *Opera Omnia.* Willis described the pylorus and compared part of it to a "chamber" or "antrum":

> **"Orificium alterum vulgo Pylorus dictus, a dextro stomachi latere cum antro capaci & longo sensim angustato, in**

foramen parvum desinit, indeque retortum in duodenum continuatur."

[The second opening, commonly called *Pylorus,* from the right side of the stomach with a wide and long, gradually narrow chamber, ends in a small hole, and then continues bending into the duodenum]

Willis here alludes to the two components of the pylorus: the *pyloric antrum* and the *pyloric canal* or *sphincter.* The pyloric antrum extends from the incisura of the lesser curvature to the pyloric sphincter, and nowadays is commonly referred to as the gastric *antrum.*

>>>>>>>>>>>>>>>>>>>>>>>>>><<<<<<<<<<<<<<<<<<<<<<<<<

w) Cardiam corpori duodecimae dorsi vertebrae adpositam esse, docet WINSLOVIUS l. c. §. 48.

w) The cardia is located at the twelfth thoracic vertebral body, so teaches Winslow, cited work, paragraph 48.

>>>>>>>>>>>>>>>>>>>>>>>>>><<<<<<<<<<<<<<<<<<<<<<<<<

In 1732, Jacques-Benigne **Winslow** wrote *Exposition Anatomique de la Structure du Corps Humain.* On page 308, paragraph 48, of his "Treatise of the Lower Abdomen" he clarified the position of the proximal opening of the stomach (or gastro-esophageal junction) and that of the cranial border of the fundus:

"**Le grosse extrémité de l'Estomac est dans l'Hypochondre gauche, pour l'ordinaire immediatement sous le Diaphragme. Cependant l'Orifice superieur de l'Estomac n'y est pas. Il est presque vis-à-vis & attenant le milieu du corps des dernieres Vertebres du Dos.**"

[The great Extremity of the Stomach is in the left Hypochondrium, and for the most part immediately under the Diaphragm. Yet the superior Orifice is not in the left Hypochondrium, but almost opposite, and very near, to the middle of the Bodies of the lowest Vertebrae of the Thorax.]

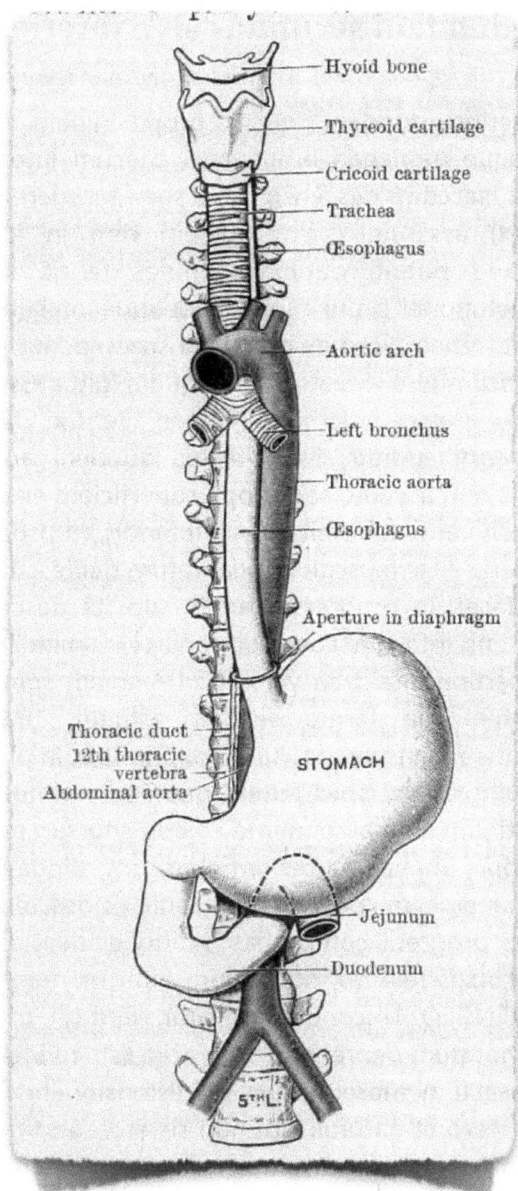

Hyoid bone

Thyreoid cartilage

Cricoid cartilage

Trachea

Œsophagus

Aortic arch

Left bronchus

Thoracic aorta

Œsophagus

Aperture in diaphragm

Thoracic duct
12th thoracic
vertebra
Abdominal aorta

STOMACH

Jejunum

Duodenum

5TH L.

From Cunningham's *Textbook of Anatomy* (1914)

§. XIII.

"De flexu duodeni secundus sive inferiori agitur"

De flexu inferiori duodeni, quem prope renem indicavimus, nonnulla quoque monenda erunt. Ad coecum intestinum, sive magnum coli saccum descensu hoc suo accedere duodenum MONROUS (x) asseruit, quam quidem rem in icone addita exprimere haud potuit, quoniam omnes ferme intestinorum tenuium convolutiones simul depinxit, ideoque intestinum coecum non delineavit, renem vero ex profunda sua sede ita protraxit, ut hepati quasi incumbere videatur. Quanquam igitur assertum suum in icone non expresserit, illud tamen veritati non prorsus contrarium deprehenditur. Nam flexus duodeni ad marginem internum renis non accedit, sed prope superficiem eius interiorem positus est. Ren vero, extremitate sua inferiori ad psoam tendens, in fossa lumbari super musculum quadratum delitescit. Nec coeco hic flexus adiacet, in progressu tamen suo ita descendit, ut ad spinam excurrens, intestino coeco opponatur, si enim linea a coeco, vel potius insertione ilei transversim ad mediam spinam ducitur, infimam partem illa ferme semper attingit, praesertim, si duodenum flatu expansum, paululum deprimitur. In prima quidem icone hoc neutiquam perspici potest, quoniam ren non apparet, si coecum intestinum cum ascendente colo in situ suo relinquitur; in altera vero icone, ubi hae partes remotae sunt, aliqua renis portio, separata pinguedine, in conspectum prodit, et descensum fere ad confinia coeci progredi comprobat. Multo evidentior est res, si mesenterio prorsus resecto, duodenum flatu multum distenditur, et, uti in GARENGEOTTI icone conspicitur, illud (y), tribus flexibus numeris suis notatum deprehenditur. Facile tamen intelligitur, hoc in casu et resectum mesenterium et distensionem flatu factam duodenum in vero et naturali situ non sistere, sed figuram potius semicircularem, quam descensum et ascensum, quem adeo exacte descripsimus, exhibere.

CHAPTER THIRTEEN

"On the Second or Inferior Flexure of the Duodenum"

Several things will also be mentioned about the inferior curvature of the duodenum, which we indicated as adjacent to the kidney. **Monro (x)** claimed that in its descent the duodenum approached the cecum or large sac of the colon, which he could not portray in the aforementioned illustration, because at the same time he depicted almost all the loops of the small intestine, and therefore could not depict the cecum. Indeed, he brought forth the kidney from its deep location, where it is seen almost lying against the liver. Notwithstanding, he then would not portray in the illustration what he stated, which, nevertheless, is found to be not entirely contrary to the truth. For the flexure of the duodenum does not reach the internal edge of the kidney but is placed close to its medial surface. In fact, the kidney, the lower extremity of which extends towards the psoas muscle, conceals itself in the lumbar fossa above the quadrate muscle. Nor does this flexure lie against the cecum, yet in its progress while running along the spine, it descends and lies near the cecum; indeed, if a line is drawn from the cecum, or rather the insertion of the ileum {ileo-cecal junction} transversely to the middle spine, this almost always makes contact with its lowest part, especially if the duodenum is expanded by gas and presses down a little. In my first drawing this cannot be seen because the kidney is not visible, unless the ascending colon in its location is detached from the cecum; in my second drawing, where these organs are removed, some portion of the kidney is brought into sight, separated of its fat, and confirms that the descent proceeds almost to the very border of the cecum. Things are much clearer when the mesentery is previously resected, and the duodenum becomes very distended with gas, as can be seen in the picture of **Garengeot (y)**, which demonstrates the three flexures. However, it is easy for one to observe, in the event that, by chance, having resected both the mesentery and caused expansion with gas, that this duodenum is not situated in its true and natural position, but rather that the descent and the ascent exhibit a semicircular shape, which we have extensively and accurately described.

98

This chapter included two footnotes "x" and "y":

>>>>>>>>>>>>>>>>>>>>>>>>>><<<<<<<<<<<<<<<<<<<<<<<<

x) loc. cit. pag. 66.

x) cited work, page 66.

>>>>>>>>>>>>>>>>>>>>>>>>>><<<<<<<<<<<<<<<<<<<<<<<<

In 1738, Alexander **Monro** wrote the article "The Description and Uses of the Intestinum Duodenum" in the 2nd edition of *Medical Essays and Observations*. On page 66, Monro said of the descending portion of the *duodenum*:

> "It descends in this cellular Sheath, till it is almost contiguous to the great Sac of the Colon, which properly is the human Caecum."

>>>>>>>>>>>>>>>>>>>>>>>>>><<<<<<<<<<<<<<<<<<<<<<<<

y) Tab. I. ad pag. 232. E. 1. 2. 3.

y) Volume 1 on page 232, letter E: 1, 2 and 3.

>>>>>>>>>>>>>>>>>>>>>>>>>><<<<<<<<<<<<<<<<<<<<<<<<

In 1728, René Croissant de **Garengeot** wrote *Splanchnologie ou L'Anatomie des Visceres*. On page 232, he gave the description to his 7th figure which highlighted the duodenum and its three curvatures (see image on page 43) The letter "E" corresponds to the duodenum with its three curves 1, 2 and 3.

Claussen's footnote mistakenly said "Tab. I" (Tabula I) whereas it should have been written as "T. I" (Tome I).

§. XIV.

"De flexu duodeni ultimo prope ieiunum pauca adduntur"

De loco autem, ad quem ascendit duodenum, antequam in ieiunum mutatur, nunc tandem disserendum est. Uti enim radix mesenterii sub mesocolo emergit, sic intestinum nostrum in eadem hac radice sinistrorsum progrediendo sursum quoque vertitur, et in ipsa mesocoli radice usque ad confinia pancreatis deducitur. Quam ob rem, eum pancreas a sinistris dextrorsum et quodammodo deorsum tendere, dixerimus § XI, flexus hic ultimus ad eam quidem altitudinem ascendere videtur, a qua iuxta pylorum exiit. Quam ob causam etiam MONROUS, qui in icone ipsa hunc ascensum non satis definire potuit, figura aliqua laterali addita, defectum hunc supplere annisus est. Quo vero distantiam utriusque partis, hoc est, principii et finis eo melius definire possemus, locum in cadavere acubus magnis notavimus, et partibus tandem remotis cognovimus, pylorum ad dextrum latus undecimae dorsi vertebrae positum esse, quam flexus duodeni superior, versus vesiculam felleam directus, parum transgreditur, finem autem duodeni, a sinistro latere primae lumborum vertebrae positum, ita ascendere, ut dum in ieiuno pergit, cartilaginem inter hanc et ultimam dorsi vertebram positam attingat; ex quo situ duodeni finem a suo principio parum distare apparet. In puella quinque annorum, macilenta, his ipsis diebus dissecta, hic ultimus flexus ad mesocolon non pertingebat, sed ex radicis mesenterii summa parte sinistrorsum sub mesocolo reflectebatur, et in ieiunum transibat.

CHAPTER FOURTEEN

"On the Last Curvature of the Duodenum near the Jejunum"

The site to which the duodenum ascends, before it becomes the jejunum, will now be discussed. Just as the root of the mesentery emerges from under the mesocolon, so also the duodenum proceeds in this very root and turns to the left and upwards, and in this same root of the mesocolon it advances all the way to the border of the pancreas. On account of this, the pancreas appears to stretch from the left to the right and somewhat downward, as we said in chapter 11, and this last flexure appears to indeed ascend toward the same height as where the pylorus came out from stomach. For this reason, even **Monro**, who in his drawing was not able to adequately define this ascent, tried to correct this defect by adding another figure on the side {i.e., his Fig. 2}. We have better defined the distance between the two parts, that is, the beginning and the end {of the duodenum}, by marking this in the cadaver with large needles, and having removed the organs we finally know that the pylorus is situated at the right side of the eleventh dorsal vertebra, that the superior flexure of the duodenum, directed towards the gallbladder, crosses over it a little, while the end of the duodenum is placed on the left side of the first lumbar vertebra. It then ascends, so that when it continues into the jejunum it reaches the location between the cartilage {i.e., intervertebral disc} and the last thoracic vertebra, where it seems that the end of the duodenum is not very far from its beginning. In a very thin five-year-old girl, who was dissected in these last days, this last flexure did not reach the mesocolon, but exited from the upper part of the mesenteric root to the left beneath the mesocolon, and then continued into the jejunum.

§. XV.

"De duodeni situ externe definiendo disseritur"

Duodeni nexum in tanta, quibus cingitur, viscerum copia difficile erui, maximeque absconditum esse eius situm ex hactenus expositis luculenter satis apparet. Cum autem et in digestionis negotio, quatenus in sanis ex voto succedit, et in variis eiusdem laesionibus accurate pervestigandis, in forensi medicina ad duodeni limites externe definiendos respiciendum sit, hanc quoque situs rationem paucis tangere e re esse videtur. Quamvis autem in medicina, foro accomodata, ex ea parte, qua vulnera diiudicat, ad laesiones duodeni vix attendendum sit, nisi simul aliae partes circumpositae gravius vunerentur, in variis tamen effusionibus humorum, post vulnera factis, in huius partis laesionem inquirere tenemur. Venena quidem oesophagum et ventriculum offendunt, interdum tamen haec cava eorum noxiam indolem eludunt, et duodeno demum inflammationis vestigia imprimuntur, ab attentio medico indaganda. Quam ob causam ad sequentia potissimum attendendum esse statuimus. Omnis dolor et tensio molesta, quae sub octava costa, quae spuriarum prima est, eiusque cartilagine profunde sentitur, indeque sub hypochondrio dextro deorsum ad renem usque descendit, duodenum potius, quam aliam partem, afficit, haec enim incommoda, si sub octava vel et septima costa in profundiori loco precipintur, atque introrsum magis, quam deorsum tendunt, in vesicula fellea latere creduntur. Sub umbilicali regione occurrentes dolores, si mox sub peritonaeo hinc inde vagantur, ieiuni convolutiones afficiunt: quod si vero fixam magis sedem, inprimis quodammodo antrorsam, servant, in flexu inferiori duodeni potius tensiones efficiunt. Sunt autem isti dolores fixi non ab anteriori solum, sed et a posteriori parte observandi, ac diiudicandi, cum retrorsum, et ad inferiorem costarum ultimarum dextri lateris ad vertebras, sensatio molesta insertionibus diaphragmatis adscribatur, quae ad duodenum, in isto loco tensum, est referenda. Facile tamen concedimus, in tanta partium vicinarum multitudine, medicum, dolores et incommoda infimi ventris disquirentem, saepius falli posse, quod quidem ex addendis sub finem consectariis pathologicis evidentissime patebit.

CHAPTER FIFTEEN

"On Defining the Position of the Duodenum from the Outside"

I discovered with difficulty the relationship of the duodenum to the great number of visceral organs by which it is surrounded, and from the previous description it is quite clear that its position is greatly hidden. However, when accurately studying both the digestive process, in so far as it is successful in a healthy person, and many of its diseases, in forensic medicine one must take into account the boundaries of the duodenum externally, and this seems to be the reason that this location is touched by few. Although in the practice of medicine, from the standpoint by which the wounds are judged, attention is hardly paid to the lesions of the duodenum, except if at the same time other surrounding organs are more seriously injured. However, in various effusions of liquids, after the wounds are made, we must search for lesions in this organ. Poisons can damage the esophagus and the stomach, but sometimes these hollow organs avoid their harmful effect, and traces of inflammation can be found in the duodenum by the attentive physician. For this reason, we must be quick to direct our attention chiefly to the following. Every pain and troublesome spasm that can be felt deep under the eighth rib and its cartilage (which is the first false rib), and then descends under the right hypochondrium to the back as far as the kidney, afflicts the duodenum more than other organs. Indeed, these troubles are believed to arise in the gallbladder if they are felt under the eighth or even the seventh rib in a deeper location, and tend more to the interior than to the back. Pain occurring under the umbilical region, if it wanders here and there just beneath the peritoneum, affects the loops of the jejunum, and if they remain more in a fixed position, especially somewhat anterior, they would produce more tension in the flexure of lower duodenum. However, when such constant pain is observed not only in the front but also in the posterior part towards the back, and at the insertion of the last two ribs to the vertebrae of the right side, such troublesome sensation is usually ascribed to the insertion of the diaphragm, though it should be attributed to the duodenum which can reach this place. We admit that due to the multitude of the neighboring organs, it is easy for a physician investigating the pain and discomfort of the lower abdomen to be frequently deceived, which will certainly be obvious at the end from the addition of the manifestly pathological implications.

....

The second half of this chapter (from "Omnis dolor..." to "saepius falli posse" was transcibed by Annibale **Omodei** in volume 38 of the *Annali Universali di Medicina* (1826) – see page 162.

Note: The phrase **"Sunt autem isti dolores . . ."** is written in Sandifort's rendition from 1778 as:

"Sunt autem isti dolores fixi non ab anteriori solum, sed et a posteriori parte observandi, ac diiudicandi, cum quum retrorsum, et ad inferiorem costarum ultimarum dextri lateris ad vertebras, sensatio molesta insertionibus diaphragmatis adscribatur, quae ad duodenum, in isto loco tensum, est referenda. "

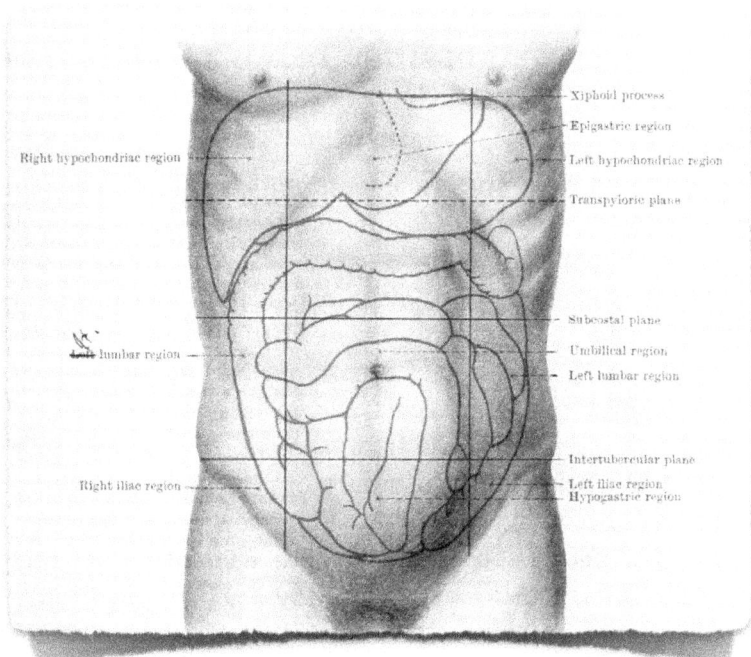

Xiphoid process
Epigastric region
Left hypochondriac region
Transpyloric plane
Subcostal plane
Umbilical region
Left lumbar region
Intertubercular plane
Left iliac region
Hypogastric region

Right hypochondriac region
Right lumbar region
Right iliac region

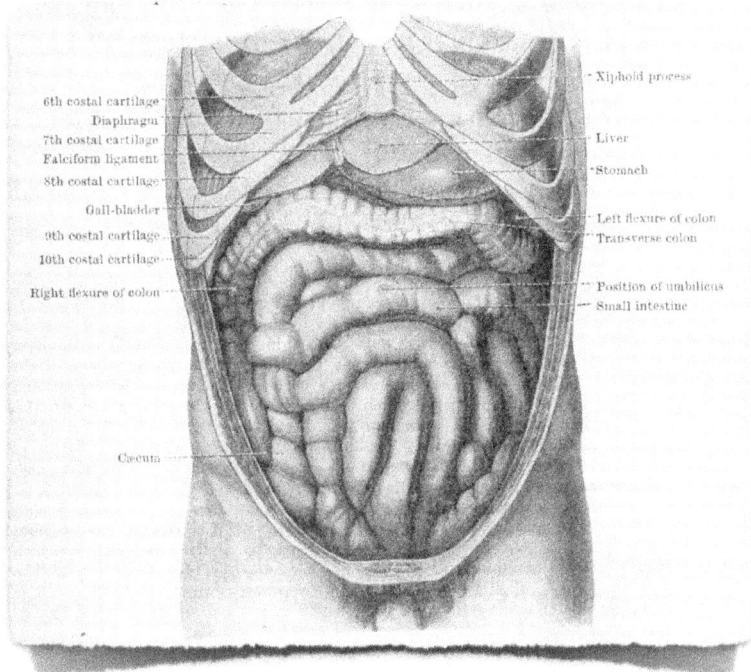

6th costal cartilage
Diaphragm
7th costal cartilage
Falciform ligament
8th costal cartilage
Gall-bladder
9th costal cartilage
10th costal cartilage
Right flexure of colon
Caecum

Xiphoid process
Liver
Stomach
Left flexure of colon
Transverse colon
Position of umbilicus
Small intestine

§. XVI.

"Duodenum situm fixum habet"

Si ergo hanc intestinorum partem, cum aliis, suis quibusvis nominibus insignitis, tubi alimentorum tractibus comparare velimus, eam prae reliquis affixam esse cognoscimus. Ventriculus quidem, in abdominis superior camera mesocolo quasi impositus, a visceribus vicinis et diaphragmate incumbente satis coercetur, in parietibus tamen suis ab omni nexu liber est, ita, ut curvatura magna, pro varia eius repletione et vicinarum partium statu, mutabilem situm recipere valeat: hinc, nisi flexus huius visceris a cardia versus pylorum insignis esset, alimenta vix assumta mox ex eo expellerentur; in duodeno autem omnes parietes affiguntur, pinguedine et contextu celluloso, ubique fere cinguntur, et membranis a peritonaeo deductis quasi alligantur. Sic quoque reliquorum intestinorum flexus a peritonaeo duplicato, in mesenterium et mesocolon expanso, suscipiuntur et margine harum membranarum tantum circumvoluuntur, quo fluctuare et ita non solum in mole sua liberius moveri, sed et eorum parietes facilius expandi contrahique possint, quam nostrum duodenum. Fixa nimirum huius intestini sedes primo quidem ratione ventriculi necessaria erat, quippe qui repletus, extremitate sua parva deorsum et paulo antrorsum tendens, duodenum, si libere fluctuaret, ex situ suo dimovere posset, id, quod, non obstante hoc nexu, interdum evenire solet; examinavimus enim aliquoties cadavera (z) in quibus per ventriculum, nimis sinistrorsum tractum, duodenum e sede sua distortum fuit, ita, ut curvatura ventriculi magna, sinistrorsum directa duodeni principium ad mediam ferme spinam diduceret. Ratione insertionis ductus cholodochi et pancreatici simili modo firma sedes requirebatur; si enim a fluctuante duodeno hi ductus varie trahi possent, illi saepius, et praefertim digestionis tempore, ubi haec cava repleta sunt, comprimerentur et liberum affluxum humorum, digestioni inservientium, non permitterent. Tandem quoque ascendens duodeni pars et finis prope ieiunum in apto situ conservanda erat, ne haec portio deorsum tracta nimis descendat, et ita iusto celeriorem progressum alimentorum, non satis elaboratorum, ad reliquum intestinorum canalem permittat.

CHAPTER SIXTEEN

"The Duodenum has a fixed position"

If we want to compare this segment of the intestine with other sections of the intestinal tract, whatever their distinctive names, we know that it is more fixed than the rest. The stomach, in fact, forced into the upper abdominal chamber by the mesocolon, is sufficiently restrained by the neighboring organs which press against it and by the diaphragm, yet its walls are free from all attachments so that its greater curvature, according to the various degrees of its fullness and of the state of neighboring organs, is able to change position. Hence, unless the curvature from the cardia to the pylorus of this organ is prominent, the food just received is soon expelled from it. On the other hand, all the walls of the duodenum are fastened within the context of the cellular and adipose tissue, and almost everywhere it is surrounded and almost tethered by the membranes derived from the peritoneum. In addition, the loops of the rest of the intestines are supported by the peritoneum, which is duplicated and stretched out as the *mesentery* and *mesocolon*, and are surrounded by the edge of these membranes. These loops can thus move in a wave-like manner and not only do they move around freely, but also their walls can easily expand and contract, just like our duodenum. The fixed position of this intestine is certainly necessary primarily because of the stomach which, when full, extends with its small end downwards and a little forward and, if it moves about freely despite this connection, could shift the duodenum away from its position, which sometimes does happen. We examined several corpses (z) in which the stomach was pulled much too far to the left, and the duodenum was moved out of its place, so that the greater curvature of the stomach dragged the beginning of the duodenum straight to the left and almost to the middle of the spine. The reason that the insertion of the bile duct and pancreatic duct is required to be in a stable position in a similar manner is that if you have a wandering duodenum these ducts will be pulled in various directions, and they often, and mainly at the time of digestion when these cavities are filled, become compressed and will not permit their free-flowing fluids to help the digestion. Finally, the ascending part of the duodenum and its end near the jejunum must be maintained in a convenient location. This part, too, does not descend to the back, so that it allows the rapid progress of the incompletely digested food into the rest of the intestinal canal.

....

This chapter included one footnote "z":

>>>>>>>>>>>>>>>>>>>>>>>>>>><<<<<<<<<<<<<<<<<<<<<<<<<

z) vid. Magnific. LUDWIGII Programma, quod observata in sectione cadaveris feminae, cuius ossa emollita erant, proponit. Lips. 1757. p. 6.

z) see *Programma: Quo observata in sectione cadaveris feminae cujus ossa emollita erant proponit* by the illustrious Ludwig, Leipzig (1757), page 6.

>>>>>>>>>>>>>>>>>>>>>>>>>>><<<<<<<<<<<<<<<<<<<<<<<<<

In 1757, Christian Gottlieb **Ludwig** (1709-1773), a professor in Leipzig, wrote the lecture *Quo Observata in Sectione Cadaveris Feminae Cuius Ossa Emollita Erant* which was presented in Leipzig on April 22, 1757. It was later included in the anthology *Disputationes ad Morborum Historiam et Curationem Facientes* edited by Haller (1758). In this presentation on the autopsy findings of a 43-year-old woman with a severe case of osteomalacia, Ludwig says:

"Ventriculi situm quoque non nihil mutatum deprehendebamus, cum non tantum extremitas ejus magna insigniter ascenderet, & curvatura magna sinistrorsum nimis versus lienem dirigeretur, extremitas quoque parva usque ad pylorum adeo angusta & callosa esset, ut cavitas digitum vix admitteret, & intestini tenuis partem mentiretur. Haec etiam ventriculi pars, magis transversim excurrens, duodeni principium ita ex situ suo deduxerat, ut proxime ad transversam zonam coli accederet, & ventriculus ipse, qui in naturali sua expansione colo proximus est, multum retractus cum mesocolo subjacens cohaereret. Duodenum post pylorum maxime ampliatum, & tandem ad vesiculae felleae latus accedens angustius inveniebatur."

[We also found that the location of the stomach was somewhat altered, not only with its large {proximal} extremity significantly raised, but also the greater curvature directed leftwards toward the spleen, and its small {distal} end at the pylorus was so narrow and hard that its lumen barely admitted a finger, and pretended to be part of the small bowel. Also, part of the stomach was oriented rather transversely and thus pulled the beginning of the duodenum

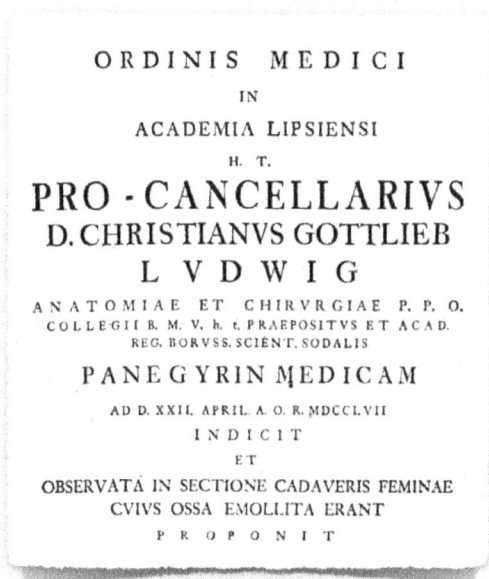

ORDINIS MEDICI

IN

ACADEMIA LIPSIENSI

H. T.

PRO · CANCELLARIVS

D. CHRISTIANVS GOTTLIEB

L V D W I G

ANATOMIAE ET CHIRVRGIAE P. P. O.
COLLEGII B. M. V. h. t. PRAEPOSITVS ET ACAD.
REG. BORVSS. SCIENT. SODALIS

PANEGYRIN MEDICAM

AD D. XXII. APRIL. A. O. R. MDCCLVII

INDICIT

ET

OBSERVATÁ IN SECTIONE CADAVERIS FEMINAE
CVIVS OSSA EMOLLITA ERANT

PROPONIT

away from its location so that it approached the transverse colon. And the same stomach, which in its normal extension is very close to the colon, was very retracted and laid against the underlying mesocolon. The duodenum beyond the pylorus was very wide, and ultimately was found approaching the narrower side of the gallbladder.]

§. XVII.

"Duodenum in situ suo liberum est"

Haec tamen intestini nostri in sede sua dispositio et arctior connexio non ea est, ut liberum alimentorum transitum impediat, sed potius permittit, ut hoc cavum, quod digestionis tempore maiorem digerendorum copiam suscipere solet, non nihil extendatur, et paulo post conveniente tunicarum robore iterum se contrahat, et ita diverso tempore variam diametrum recipiat. Membranae enim ligamentosae, quae a variis peritonaei partibus eo deducuntur, non ubique tunicae musculosae firmiter adhaerent, in macilentis quoque cellulosa substantia, laxius disposita intersternitur, quae efficit, ut expansioni, ab alimentis introductis factae, undiquaque cedere queat. In cadaveribus, modica pinguedine praeditis, hoc melius cognoscitur, in his enim non tantum descensum, a latere vesiculae felleae et sub dextra mesocoli parte, sed et ascensum in mesenterii radice, copiosa pinguedo cingit, porro quoque in tribus flexibus, ubi membrana a peritonaeo deducta, ad musculosam tunicam agglutinata deprehenditur, vicinae partes pinguedine apposita molliter habentur. In subiectis obesis nimia fere haec videtur pinguedinis copia, quae intestinum nostrum partim mole sua comprimit, partim eius ambitum relaxando insignem extensionem permittit, ideoque diversa obesorum incommoda inferius commemoranda efficit. In omni tamen huius intestini statu, tanta eidem manet libertas, ut tensionibus et vicinarum partium pressui non adeo obnoxium sit, sed alimentorum susceptorum ulteriorem praeparationem commode praestare, et motum peristalticum rite exercere possit.

CHAPTER SEVENTEEN

"The Duodenum is free from restraint in its location"

The arrangement and connection of the duodenum in its location is not very tight, since this would impede rather than allow the free passage of food, as this cavity, which usually takes on a greater amount of digested food, is sometimes dilated and shortly thereafter contracts itself again with appropriate force of its tunics, and thus presents a variety of diameters at different times. Indeed, the ligamentous membranes, which are derived from the various parts of the peritoneum and do not firmly adhere to the muscular coats, are stretched out in the loosely arranged thin and cellular substance, which ensures that the expansions made by the introduced food can yield on all sides. This is best seen in cadavers which possess a moderate amount of fat, in which indeed not only the descent from the side of the gallbladder and under the right mesocolon but also the ascent within the root of the mesentery is surrounded by abundant fat. Also, along the three curvatures, where the membrane derived from the peritoneum is observed adherent to the muscular coat, their neighboring organs have soft adipose tissue placed nearby. In obese persons there is an almost excessive amount of fat, which partly compresses our intestine with its mass, and partly allows its area to relax by remarkable expansion, and therefore creates various harmful effects in the obese as mentioned below. However, in every situation of this intestine, it has a lot of freedom since it is not subject to great tensions and pressure by neighboring organs, but comfortably performs further preparation of the food received, and can duly exercise its peristaltic motion.

§. XVIII.

"Progressus duodeni maxime incurvatus est"

Omnis alimentorum tubus flexuosa via progreditur, et non tantum ideo continuis gyris circumvoluitur, quo ratione longitudinis in situ commodo retineatur, sed quo peristalticus etiam motus, in ea potissimum huius canalis parte, quae sub umbilico posita est, nimis celer futurus, anfractuosa hac via paululum refrenaretur. Quod quidem assertum dilucide comprobatur, si ventriculi, intestini ieiuni et zonae coli flexus consideramus, simulque progressum alimentorum vel digerendorum, vel post aliqualem praeparationem factam chylum exhibentium, vel tandem exsuccorum et in faeces mutatorum inter se comparamus, et inde partiales intestinorum usus colligimus. Quod si vero ambitum duodeni cum nunc dictis partibus conferamus, flexum eius integrum a pylorous quead ieiunum adeo incurvatum deprehendimus, ut, praeter aptum situm remoram alimentorum in hac parte digerendorum perspiciamus. Chymum ex ventriculo egredientem et ipsa extremitas parva sursum et introrsum flexa, et prima, quae ad pylorum est, duodeni curvatura non satis retineret, nisi simul pylori angustia et introrsum productum iugum valvulosum celeriorem egressum impediret: postquam vero hunc anfractum transit, massa, ulterius digerenda, in descendente parte subito defertur, rugae enim valvulosae, ex nervea ex villosa tunica formatae, in duodeno exiguae sunt, nec adeo prominent, ut in sequenti ieiuno. Remora autem praecipue in media ferme duodeni parte iniicienda erat, ubi nimirum liquores digestionem adiuvantes admiscentur, hoc igitur in loco ascensus duodeni versus ieiuni principium oblique sursum absolvitur, per quem alimenta quoque cum quadam difficultate sursum moventur, et mutationem mox describendam melius subeunt. Tandem quoque ultimo flexu, prope ieiunum observato, alimentis adscendentibus et versus hoc intestinum propulsis obex ponitur, in primis cum rugae valvulosae, hactenus depressae, circa ieiuni initium quodammodo assurgere observentur.

CHAPTER EIGHTEEN

"The course of the Duodenum is very tortuous"

The entire digestive tube proceeds in a tortuous fashion, and thus not only does it twist around in a series of continuous loops, for which its entire length is maintained in a convenient location, but its peristaltic movement, which is faster in the segment of this canal which is situated under the umbilicus, is somewhat slowed down by all its windings. This assertion is clearly demonstrated if we consider the flexures of the stomach, jejunum, and colon, and compare the progress of food or digestion, or the chyle produced after a certain preparation is made, or lastly the fluids transformed into feces. We can then infer the function of the various parts of the intestine. But if we consider the area of the duodenum with the organs just mentioned, we discover that its entire curvature from the pylorus all the way to the jejunum is so greatly curved, that, in addition to a fixed position we notice a delay of the digesting food in this organ. Both the small end of the stomach, which is bent upward and inward, and the first curvature of the duodenum, located at the pylorus, do not delay enough the chymus that comes out of the stomach, unless at the same time the narrowness of the pylorus and the internal archway made by the valve impede a faster exit. After it passes through this bend, the mass of food, further digested, is immediately carried to the descending portion. The valve-like rugae formed by the submucosal and mucosal layers are very small in duodenum and are not so prominent as in the following jejunum. But a delay is caused almost immediately in the middle of the duodenum, where the fluids assisting in the digestion are mixed. Then there is the oblique ascent of the duodenum that ends up towards the beginning of the jejunum, through which food is also transported upwards with some difficulty and undergoes more of the changes previously described. Finally, at the last curvature near the jejunum, an obstacle is placed for the food ascending and propelled toward this intestine by the valve-like rugae, which up to this point were low, and are observed to somewhat increase in height near the beginning of the jejunum.

§. XIX.

"Causae digestionis in duodeno concurrentes brevibus indicantur"

Alimenta in ventriculo non nihil soluta et in pultaceam massam reducta, ex eo in duodenum transferuntur. In hoc cavo omnes digestionis causae, quae in ventriculo adfuerunt, quoque deprehenduntur, calor scilicet partibus internis conveniens, motus parietum peristalticus, et agitatio a vicinorum viscerum pressu motuque diaphragmatis et musculorum abdominalium respiratione concurrentium. Affluit etiam duodenalis liquor, licet ille copia sua cum gastrico neutiquam comparandus sit. Praeter has et alias, hic non fusius recensendas digestionis, causas, liquores praestantissimi, resolutionem et praeparationem alimentorum digerendorum efficientes, bilis nimirum hepatica et cystica, nec non succus pancreaticus, in nostrum intestinum ingeruntur. Instituti ratio non permittit, ut cuncta, quae de horum liquorum ortu, indole et natura proponi possunt, repetamus, cum ex physiologicis doctrinis satis pateant. Sufficiat itaque in genere monere, bilem praecipue particulas oleoso-salinas, et inter se et cum reliquis componentibus partibus aquoso terrestribus, ita commixtas habere, ut vere saponaceus, hoc est, potenter resolvens, liquor inveniatur, et qua proprietate viscidas et pingues alimentorum moleculas penetrat, easque cum aquosis partibus ita coniungit, ut ab orificiis vasorum, chylum resorbentium, facile suscipi, et per tenuissimos glandularum meseraicarum maeandros transire possint. Idem fere praestat succus pancreaticus, ex ductu Wirsungiano affluens, qui salivae simillimus non solum resolvit, sed spissam et summe amaram bilem cysticam, in valida digestione expressam, diluit, et particulis alimentorum intimius resolvendis magis magisque applicat. Quoniam autem et alimenta non nunquam acriora, et liquores resolventes stimulo suo ea, quae in duodeno elaboranda sunt, nimis cito propellerent, mucus intestinalis, qui ex glandulis, in duodeno copiosus, quam alibi, positis, excernitur, nimiam irritationem infringit et placidam digestionem in duodeno sistit.

CHAPTER NINETEEN

"The simultaneously occurring causes of digestion in the Duodenum are briefly discussed"

Food is converted to a somewhat loose and pultaceous mass within the stomach from where it is transferred to the duodenum. In this cavity all the mechanisms for digestion which occur in the stomach are also detected, such as the heat obviously generated by the internal organs, the peristaltic movements of the walls, the agitation by nearby abdominal viscera as well as the combined pressure and respiratory movement of the diaphragm and abdominal muscles. The duodenal secretions also flow in, though this is by no means comparable in quantity to the gastric juice. Besides these and other mechanisms of digestion, which have not been fully accounted for, outstanding fluids that are efficient in the resolution and digestion of the food, such as the hepatic and gallbladder bile, as well as the pancreatic juice, are poured into the duodenum. The aim of our investigation does not allow us to repeat everything about the origin, character and nature of these body fluids, since these are sufficiently revealed from the teachings of physiology. Therefore, in general it is sufficient to recall that especially the particles of the oily-salty bile, both by themselves and with the rest of the earthly watery parts, when mingled will form a truly saponaceous fluid, that is, a liquid with the power to loosen. Due to this property it penetrates the viscid and fatty molecules of food, and thus mixes these with the watery parts, so that it is readily accepted by the openings of the reabsorbing lymph and blood vessels, and can pass through the tiny windings of the mesenteric glands. The pancreatic juice provides almost the same, flowing from the duct of Wirsung, which very much like the saliva not only loosens but also dilutes the dense and very bitter bile, secreted during a healthy digestion, and it is added more intimately to the particles of food which become more and more loosened. However, since both the sometimes-bitter food and the loosened liquids, which are elaborated upon in the duodenum, are propelled very quickly, the intestinal mucus secreted from the glands {i.e., Brunner's glands}, which are more abundant in the duodenum than in other places, weakens the excessive irritation and brings about a calm digestion in the duodenum.

....

This chapter introduces some elements of physiology of digestion, many of which are repeated from previous Authors, such as Vesalius. Some of these notions would continue to to be taught in 19th century textbooks such as in Blumenbach's *The Institutions of Physiology* (1820), the author of which most definitely had read Claussen's thesis. In fact, we find in this book the relevant passage wherein the factors involved in the digestive process of the stomach are discussed:

> **"The other aids commonly enumerated, are the pressure on the stomach from the alternate motion of the abdomen, and the high temperature maintained in the stomach by the quantity of blood in the neighbouring viscera and blood-vessels, which temperature was at one time supposed to be of such importance, that the word coction was synonymous with digestion."**

Indeed, the Latin term "coquo" means *to cook by heat*, and "concoquo" means *to boil together* or *to digest*. And consequently, the term concoction was often used to describe the processing of food in the stomach. Another meaning for "concoquo" is *to bear* or *endure* and, appropriately, *"to stomach."*

§. XX.

"Usus duodeni generalis declaratur"

Omnia itaque, quae de situ et nexu intestini nostri, ex repetita et accurata eius contemplatione, cognovimus, nos ad physiologica quaedam consectaria deduxerunt, ex quibus coniunctis, usum duodeni generalem nunc brevibus definire tentabimus. Primarium digestionis organon ventriculus dicitur, quippe, qui alimenta, masticatione emollita et saliva remixta suscipit, retinet, movet, resolvit, et in pultaceam massam, quam chymum dicimus, mutat. Quemadmodum vero praeparatio in ore, et apta salivae admixtio, digestionem a ventriculo praestandam incipit, sic ventriculus illam continuat, non vero penitus absolvit; chymus enim e ventriculo egrediens, chylum dimittere haud posset, nisi actio duodeni massam digerendam ulterius solveret, et ad chyli separationem disponeret, Praestant hoc liquores affluentes § XIX, intimamque particularum nutrientium evolutionem sistunt, sub qua illae, ut peregrinae, in humorum corporis humani naturam sensim paulatimque convertuntur. Haec functio, quae primaria et vera digestio iure suo dici meretur, non absolvi posset, nisi duodenum, cuius situm et nexum declaravimus, in fixa sede haereret, § XVI. nisi illud simul alimentis progredientibus liberim motum concederet, § XVII. nisi tandem progressu suo maxime flexuoso et incurvato alimenta ad tempus retineret, § XVIII. et ita exquisitissimam digerendorum elaborationem praestaret. Ex quibus collectis facile intelligiitur, duodenum intestinum esse illam tenuium intestinorum partem, quae alimenta semidigesta ex ventriculo recipit, ad tempus retinet, lentius movet, digerentium liquorum, bilis et pancreatici succi, admixtionem absolvit, et ita alimenta vere digesta ad ieiunum propellit. § IX.

CHAPTER TWENTY

"The general function of the Duodenum"

Thus, everything that we now know about the position and relation of the duodenum, from repeated and accurate observations, has led us to certain physiological inferences, from which we will now attempt to briefly define the general function of the duodenum. The stomach is said to be the primary organ of digestion. In fact, it receives, retains, moves, and dissolves the food that was softened after being chewed and mixed with saliva, and transforms it into a pultaceous mass, which we call *chyme*. Indeed, the preparation starts in the mouth with the addition of saliva, and the digestion begins to be performed by the stomach, and thus the stomach continues but does not completely finish it. The chyme exiting out of the stomach cannot release the chyle, unless the action of the duodenum further digests the mass and arranges a separation into chyle. The incoming fluids, which are discussed in chapter 19, aid the complete transformation of the nutritive particles, so that those foreign to the body are slowly and gradually transformed into natural fluids of the human body. This function, which deserves to be called in its own right the first and true *digestion*, cannot be completed unless the duodenum, the location and the connection of which have already been explained, stays in a fixed place (see chapter 16), and, at the same time, allows free movement to the progressing food (see chapter 17) and, finally, retains the food for a while in its sinuous and curved course (see chapter 18). In this manner it provides an exquisite elaboration of the digested food. From these conclusions it is easy to understand that the duodenum is that part of small intestine which receives the semidigested food from the stomach, retains it for a time, slowly moves it, completes the admixture of digestive fluids, bile and pancreatic juice, and thereafter propels the truly digested food into the jejunum (see chapter 9).

§. XXI.

"Duodenum a partibus vicinis, morbo affectis, laeditur"

Usus duodeni intestini generalis, ex hactenus propositis stabilitus, in sano corpore egregius est: quanta vero ille in successiva humorum mutatione adfert commoda, tanta quoque oriuntur vitae incommoda, si vel partes duodeno vicinae male se habent, vel ipsam in fabrica sua, aut in nexu et situ quodam modo laeditur. Quam quidem tractationem pathologicam, ad finem properantes, vix digne satis evolvere, sed tantum paucis adumbrare suscipimus. Ventriculum, e sede sua non nihil dimotum, intestini nostri initium sinistrorsum ducere § XVI declaravimus. Sub qua sitos mutatione, ut plurimum spasmi oriuntur, quibus una alterave pars angustior redditur, vicina paulo post sese expandente, ex quibus diametri mutationibus varia incommoda generantur, inprimis, si vel cellulosa tela, nerveae potissimum tunicae, vel glandulae mucosae callosae et scirrhosae redduntur. Ab hepate quoque, mole et pndere aucto, duodenum nostrum eiusque primum flexum comprimi posse, similemque vim a fellea vesicula, calculi distenta, applicari dubium non est. Ab adscendente et duodeno incumbente coli parte hoc incommodum raro evenire, supra monuimus, § VII, cum ab hoc latere remora seybalorum, vicini partibus molesta, in coeco potius observetur, oblique tamen adscendens duodeni pars, in scirrhosa glandularum mesaraicarum labe a compressione libera non est; ex quibus omnibus patet, quomodo alimentorum in duodeno remora, quae in statu naturali adeo necessaria est, per causas morbosas molesta fieri, et digestionem laedere soleat.

CHAPTER TWENTY-ONE

"The Duodenum is damaged by neighboring organs affected by disease"

The general function of the duodenum, as established from what we have proposed up to this point, is admirable in a healthy body: it contributes so much to the appropriate successive transformation of the fluids that if the organs near the duodenum are diseased or the duodenum is damaged in some way in its structure, position or relations, these make life uncomfortable. It is difficult, in fact, to adequately discuss the pathological changes in a hasty manner, but we will undertake to outline a few. We have shown how the stomach, when moved away from its normal position shifts the beginning of the duodenum towards the left (see chapter 16). With this change in position, many spasms arise, in which one or another section becomes narrower, and proximal to this the duodenum expands itself, and this change in diameter generates various injuries, especially if either the cellular layer {mucosa}, more so the nervous layer {submucosa}, or the mucous glands {of Brunner} are rendered hard and fibrotic. There is no doubt that the duodenum and its first flexure can be compressed by a liver that is increased in size and mass, and likewise by a gallbladder distended by calculi. We previously pointed out that this problem rarely occurs due to the ascending colon pressing upon the duodenum (see chapter 7), because from this side the delayed lumpy stools, which damage neighboring organs, are preferably observed in the cecum. However, the ascending portion of the duodenum is not free from compression by the fibrosis of the mesenteric lymph nodes. From all this it is clearly evident that the delay of food in the duodenum, which is so necessary in the natural state, if caused by diseased conditions tends to harm the digestion.

§. XXII.

"Duodenum nimium ampliatur"

In statu naturali intestinum nostrum tempore digestionis maiorem alimentorum copiam suscipere, paullulum retinere, inde haud raro iusto maiorem in modum extendi, et a quibusdam ventriculum succenturiatum appellari (a), observatione constat. Robore tamen muscularis tunicae illaeso; post digestionem hoc in loco absolutam, blanda huius partis contractio succedit, inprimis cum bilis affluens stimulo suo concurrat, et flatus, distensiones efficientes, discutiat ac promoveat. Alimenta tamen digestioni resistentia, et flatulenta copiose ingesta, et bilis inertia per varias causas inducta, praeternaturalis distensionis causae sunt praecipuae, hinc digestion omnino laeditur, et cruda nec satis subacta propelluntur. Cum itaque duodenum digestionem, in ventriculo nondum absolutam, ulterius continuare, et intimam alimentorum resolutionem in cavo suo praestare debat, multa vitia primae concotionis relinquit, quae in continuato intestinorum tractu non satis corriguntur, sed in reliquis etiam chylificationis, sanguificationis et secretionis, immo nutritionis ultimae functionibus haud leves morbos producunt. Hi quidem in obesis et voracibus praecipue cognoscuntur, in quibus alimentorum elaboratio, et inde pendens nutritio, per plures annos ex voto succedit, tandem vero ita laeditur, ut molestissima asthmata, multis symptomatibus stipata, eos excrucient. Pinguedo enim, circa intestinum nostrum collecta, § VIII si modica est, liberum eius motum in digestione egregie iuvat, § XVII nimis autem accumulata tunicas relaxat, et eorum naturale robur infringit, quo facto motus peristalticus libere non succedit, sed spasticae contractiones, accedente inprimis flatuum copia, urgent, totumque digestionis negotium, in duodeno praecipue perficiendum, § XX turbant.

CHAPTER TWENTY-TWO

"The Duodenum can be excessively widened"

It is evident that in the natural state the duodenum takes a longer time to digest food, detains it a little, and not uncommonly is stretched in a wider manner, and for this reason some have called it the *succenturiate stomach* (**a**). However, with the vigorous strength of its muscular coat, after the digestion is completed in this place, a mild contraction of this organ will follow, especially when the inflowing bile with its stimulus, and gas, by causing distention, disperses and pushes it forward. However, food resistant to digestion, abundantly ingested gas, and sluggish bile induced by various causes are the main reasons for an unnatural distention, which can absolutely affect the digestion, and cause raw and insufficiently prepared food to be propelled forward. Therefore, since the duodenum must continue the digestion of food not yet completed in the stomach, and supply within its cavity the most complete dissolving of the food, it would leave unfinished many defects of the first concoction, which cannot be corrected in the following intestinal tract. Instead, the nutritional function of the previous will produce rather serious malfunctions in the remaining chylification, sanguification and secretion. These are especially known in the obese and voracious individuals, in whom after the suspension of feeding for several years because of a vow, the elaboration of food is ultimately so damaged that troublesome breathing difficulties full of many symptoms distress them. The adipose tissue around the duodenum (see chapter 8), if it is modest in amount, aids in its freedom of movement during digestion (see chapter 17), but if excessive it relaxes too much the coats and infringes upon their natural strength, and consequently the peristaltic motion does not occur freely, and especially with the addition of much gas, the spastic contractions strongly compress and disturb the whole process of digestion, which is chiefly completed in the duodenum (see chapter 20).

This chapter included one footnote:

>>>>>>>>>>>>>>>>>>>>>>>>>><<<<<<<<<<<<<<<<<<<<<<<<<

a) MONROUS l. c. pag. 71. seq. miramur itaque, cur BOERHAAVIUS in Institi. med. §. 96. et SENACUS l'anatomie de Heister, à Paris 1755. 8vo maj. p. 129. cavitatem duodeni reliquis intestinis angustiorem et minorem dicant, cum tamen naturalem intestini huius statum describant. Auctores, qui naturalem amplitudinem concedunt, HALLERUS in notis ad § cit. BOERHAAVII recenset.

a) Monro, cited work, page 71; and we marvel at the following, where Boerhaave in his *Institutionis Medicus*, chapter 96, and Sénac in *L'Anatomie d'Heister*, Paris (1755), octavo major, page 129, said that the duodenal lumen is narrower and smaller than the rest of the intestine, when in fact they describe this intestine in its natural state. Haller reviewed the Authors who agreed with the natural wideness in his annotation to the cited chapter {#96} of Boerhaave.

>>>>>>>>>>>>>>>>>>>>>>>>>><<<<<<<<<<<<<<<<<<<<<<<<<

In 1738, Alexander **Monro** wrote the article "The Description and Uses of the Intestinum Duodenum" in the 2nd edition of *Medical Essays and Observations*. On page 71 he wrote:
> "It is no Wonder that this Intestine is frequently found of so much larger Diameter than the other Guts, as to be called Ventriculus Succenturiatus by several Authors."

But Claussen complains about two authors, Boerhaave and Sénac, who stated the exact opposite in regard to the amplitude of the duodeum.

In 1708, **Boerhaave** wrote the book *Institutiones Medicae*. In its chapter 96, he said of the *duodenum*:
> "Intestino duodeno propria est rectitudo, sine valvulis fere, angustia, . . ."

[The *duodenum* itself is straight, usually without valves, narrow, . . .]

In 1735 (not 1755 as Claussen says), Jean-Baptiste de **Sénac** (1693-1770) wrote the 2nd edition of his book *L'Anatomie d'Heister, avec des essais de physique sur l'usage des parties du corps humain, & sur le*

Jean-Baptiste Sénac

méchanisme de leurs mouvemens. This was basically a French translation of the book *Compendium Anatomicum* (1717-1792) by the German anatomist, surgeon and botanist Lorenz **Heister**. On page 129 of volume one, Sénac wrote:

"Le duodenum qui tire son nom de douze pouces de longueur qu'on lui donne ordinairement, il commence au pilore: il est d'abord perpendiculaire de haut en bas, il s'étend ensuite horisontalement du coté droit de l'andomen vers le rein gauche, à trois ou quatre doigts du pilore il reçoit "l'ouverture du conduit coledoque & du pancréatique, qui y conduissent la bile & le suc du pancréas, ses tuniques sont plus épaisses que dans le reste des intestins, mais la cavité est un peu plus petite, son artére vient de la coeliaque, sa veine se rend à la veine-porte comme celles des autres intestins, les glandes y sont fort nombreuses, on les appelle glandes de Brunner, elles filtrent une liqueur subtile."

Similar wording can be found on page 128 of the 1st edition of Sénac's book from 1724. The 3rd edition fom 1753 had the same wording, but this time it was on pages 234-5.

Below is the translation as it originally appeared in the English version of Heister's book called *A Compendium of Anatomy* from 1721:

"The first of these is call'd Duodenum, because 'tis twelve Inches long; it takes its Rise from the Pylorus, and runs first perpendicularly, then horizontally, from the right Part of the Abdomen towards the left Kidney. At the Distance of three or four Fingers from the Pylorus, two Canals open into it call'd Ductus Cholidochus, and Ductus Pancreaticus; from the one, it receives the Bile, from the other, the Pancreatic Juice. The Thickness of its Coats is greater than that of the rest of the small Guts, but its Cavity is commonly less. It receives an Artery from the Coeliac; and like the rest of the Guts, it has its Vein from the Porta. Brunnerus's Glands are very numerous here for the Secretion of a thin Liquor."

It was in 1717 that Lorenz **Heister** (1683-1758) issued the 1st edition of his *Compendium Anatomicum*. When listing facts about the duodenum he made no mention of its lumen in this particular edition. It was only in the 2nd edition from 1719 that Heister said:

"Tunicarum eius crassities major, quam reliquorum tenuium; sed cavitas fere minor."

[The thickness of its coats is greater than that of the rest of the small intestines; but its lumen is generally less.]

Thereafter, Heister's book would be re-issued many times up until the 9th edition of 1792. However, it was in the 3rd edition from 1730 that this specific wording changed:

"Tunicarum ejus crassities & cavitas, major, quam reliquorum tenujum."
[The thickness of its coats and the lumen are greater than the rest of the small intestines.]

And in the 4th edition of 1736, he said:

"Tunicas habet crassiores, & cavum maius, quam reliqua tenuia."
[It has thicker coats, and a larger lumen, than the rest of the small intestines.]

So, to be fair, before Sénac's 1735 translation into French, Heister had already corrected his assertion regarding the duodenum's lumen.

The Italian edition *Compendio Anatomico di Lorenzo Heister* from 1772 was translated from Heister's 4th edition and gave him justice by saying:

"Ha le tonache più grosse, e la cavità maggiore degli altri intestini tenui."
[It has thicker coats, and a larger lumen than the other small intestines.]

Claussen ends this footnote with a reference to the writings of Albrecht **Haller**. This regards the lectures of Boerhaave with footnotes and commentaries by Haller himself called *Praelectiones Academicae* from 1740. Claussen is referring to Haller's footnote "2" in chapter 96 wherein he discusses the supposed "narrowness" of the *duodenum*. Haller stated:

"Angustia, in Text.) Imo vero omnibus tenuibus intestinis latius est. VESAL. L. V p. 499. ed. 1543. SANTOR. l. c. WINSL. IV 119. MONROO l. c. p. 71. BRUNNER. de Duoden. VESLING. Syntagm. anat. c. III."
[*Narrowness*: On the contrary, it is, in fact, the widest of all the small intestines. **Vesalius**, Book 5, page 499, edition from 1543. **Santorini**, cited work. **Winslow**, book 4, paragraph 119. **Monro**, cited work, page 71. **Brunner**, *De Duodeno*. **Vesling**, *Syntagma Anatomicum*, chapter 3.]

The above references cited by Haller allow us to review what was written by those Authors who, according to Haller, agreed that the duodenum was wider than the remaining small bowel.

(1) First, we have Andreas **Vesalius** from his *Humani Corporis Fabrica* (1543). On page 499 of Book 5, he wrote the passage entitled "**Gracilis intestini forma**" [The shape of the thin intestine]:

> "**Ac tales gracilis intestini ter nomen emutantis, ductus situsque sunt. Forma autem undecunque constat simili, ad amussim videlicet tereti, & omni ex parte aeque ampla, nisi interdum (ut prius quoque innvebam) qua dorso proxime committitur, ieiunum que incipit, ac ad insertionem vasis bilem ad intestina deferentis, amplius observetur.**"

> [The course and position of the small intestines are such that it changes its name three times. Its shape is similar throughout, that is to say exactly rounded, and in all of its parts it is equally wide, though sometimes (as I hinted earlier) it is observed to be wider near the back where it commences, where the jejunum begins, and also at the insertion of the duct that carries the bile to the intestines.]

(2) In 1724, Giovanni Domenico **Santorini** wrote the book *Observationes Anatomicae*. In chapter IX, page 167, Santorini made the following observation of the *duodenum*:

> "**Latior caeteris tenuibus intestini duodeni tubus est.**"

> [The duodenal tube is wider than the rest of the small intestines.]

(3) In 1732, Jacques-Benigne **Winslow** wrote *Exposition Anatomique de la Structure du Corps Humain*. On page 512, paragraph 119 of the "Treatise of the Lower Abdomen," he wrote:

> "**Cet Intestin est ordinairement le plus ample, quoique le plus court des Intestins Grêles. Il est environné de plus de Tissu Cellulaire que les autres, surtout dans son Etui triangulaire, oú el n'est pas totalement environné d'une Tunique Membraneuse commme les autres, & par consequent plus susceptible de dilatation par les matieres qui seroient arrêtées dans sa cavité.**"

[This intestine is ordinarily the widest, albeit the shortest, of the small intestines. It is surrounded by more cellular tissue than the others, especially within its triangular case where it is not entirely surrounded by a membranous coat {peritoneum} like the others, and consequently it is more susceptible to dilatation by the substance which might otherwise halt within its lumen.]

(4) In 1738, Alexander **Monro** wrote the article "The Description and Uses of the Intestinum Duodenum" in the 2nd edition of *Medical Essays and Observations*. On page 71 he wrote:

"... it is no Wonder that this Intestine is frequently found of so much larger Diameter than the other Guts, as to be called Ventriculus Succenturiatus by several Authors."

(5) In 1688, **Brunner** wrote the monograph *De Glandulis in Intestino Duodeno hominis detectis* wherein he describes the glands that will be named after him. He begins by describing the *duodenum*. After citing Vesling's statement (see below) he says:

"Caeterum mirabilem init in carina corporis flexuram, per quam dum revolvuntur contenta, tardius & lente progredi videntur; faciliusque moram inibi nectunt, dum per pylorum, ceu angiportum in amplum duodeni spatium incidunt, ubi motus diffunditur, non aliter atque aqua per rivulum influens in amplum spatium, stagnum facere solet."

[In addition, in the keel of the body begins a remarkable curvature, through which the contents as they revolve seem to progress slowly and calmly; and more easily produce a delay there, while they drop through the pylorus, like a narrow street, into the ample space of the duodenum, where the movement is spread out: and not unlike the water of a stream that flows into a wide space where it tends to form a lake.]

(6) In 1641, Johann **Vesling** (1598-1649) published the 1st edition of his *Syntagma Anatomicum* (see detail of its frontispiece on page 131 which illustrates an anatomic dissection lecture). When discussing the *duodenum* in chapter 3, page 30, Vesling says:

"Laxitate interdum, & amplitudine insigne praeditum est, ut majori ventriculo minorem alterum addat, felleo meatu longius tunc protracto."

[Sometimes it is provided with exceptional spaciousness and width that it adds a smaller second stomach to the larger one, in which case the bile duct is prolonged further.]

. . . .

N.B.: "Sanguification" – an old term for the natural function of the body by which the chyle is transformed into blood. Galen (129 AD–200 AD) had proposed the theory of sanguification whereby after the chyle was taken up by the veins of the intestines, it was carried to the liver where it was changed into red blood. This theory was abandoned once Pecquet discovered the principal lacteal vessel or thoracic duct which terminated into the left subclavian vein.

IOANNIS VESLINGII
MINDANI

SYNTAGMA
ANATOMICUM

§. XXIII.

"Duodenum nimis extensum partes vicinas premit"

Duodeni autem praeternaturalis expansio non tantum functiones eius laedit, sed etiam partes vicinas premendo, haud vulgares molestias inducit. Quamvis enim vesicula fellea, ab expanso duodeno pressa, progressum bilis cysticae in valida digestione necessariae, iuvet, ductus tamen cholodochus, fine suo inter tunicas prorepens, simili modo pressus aequabilem huius liquoris saponacei ingressum praestare haud valet. Quod si vero, omnibus resistentiis superatis, bilis vehementius irruat, tensione flexus duodeni primi prope pylorum, et tertii prope ieiunum conspicui aucta, insignis spasmus oritur, qui anxietates et dolores efficit, et haud raro ita intenditur, ut alimenta semidigesta partim in ventriculum repellantur, et aucto malo saepius per vomitum reiiciantur, partim versus tenuia intestina protrudantur, ideoque accedente alimentorum ac bilis acrimonia, choleram, dirum illum morbum, producant, qui in duodeno excitatus sursum deorsumque progreditur, et post spasticos inflammatorios-que effectus, licet interdum superatos, insignem tamen canalis alimentorum debilitatem relinquit. Haec incommoda et eam ob causam augentur, quod pancreas ex parte duodenali nimium premitur, et ita liquorem blandum, salivae similem, quo digestio adiuvatur, bilisque acrimonia corrigitur ac temperatur, libere non exhibet. Non semper quidem tam violenta expansio, spasticis motibus stipata, animadvertitur, sed et ea, quae in retardato digestionis negotio evenit, molesta est; cum enim a flexu renali adscendens duodeni pars, in mesenterii radice posita, vasis lacteis ex glandulis concurrentibus, et ad cisternam chyli properantibus, vicina sit, nullum restat dubium, et elaborationem et progressum chyli, hac duodeni expansione, multum infringi.

CHAPTER TWENTY-THREE

"The excessively dilated Duodenum presses against nearby organs"

The abnormal expansion of the duodenum interferes not only with its functions but, by pressing on nearby organs, also leads to uncommon disturbances. Even though compressed by the expanded duodenum, the gallbladder assists in the the progress of the bile necessary for a proper digestion; the choledochus, on the other hand, with its distal end creeping forth between the layers {of the duodenal wall}, in a similar manner cannot guarantee entry of its soapy fluids while under the same pressure. But if, in fact, with all resistances overcome, the bile forcefully rushes in and the tension becomes distinctly increased in the first curvature of the duodenum near the pylorus and the third near the jejunum, and a major spasm arises that produces anxiety and pain. It is not at all rarely recognized that the semidigested food is partly repelled back into the stomach, which often rejects the great evil through vomit, and partly is pushed forward towards the small intestine, where the addition of food and the bitter bile produce that horrible disease known as *colic*, which in the excited duodenum proceeds upwards and backwards, and after the spastic and inflammatory effect, even though at times it is overcome, it leaves a greatly debilitated alimentary tract. These maladies are increased both for this reason and because the pancreas is excessively compressed by the duodenal organ, so that its bland fluid which, similar to saliva, helps the digestion and corrects and dampens the bitterness of the bile, is not freely produced. It should be noted that the dilatation is not always so violent and full of crampy movements, but it may still be troublesome because a delay in the digestive process may occur. In fact, since the portion of the duodenum ascending from the renal flexure is situated within the root of the mesentery, and thus in the vicinity of the lacteal vessels travelling from the glands and straight to the chyliferous duct / cistern, there is no doubt that an expansion of the duodenum significantly impairs both the production and the progress of the chyle.

§. XXIV.

"Duodenum in nimia expansione vasa sanguifera comprimit"

Sanguinem, in suo per vasa abdominalis motu, multis modis retardari, atque ita variis morbis, chronicis praefertim, ansam suppeditare, res est certissima. Vena portae quidem peculiare et a cava prorsus distinctum systema sistis, quo regressus sanguinis per abdomen ad cor eo felicius absolvi possit; hoc tamen non obstante multa, in hac transuenda via, occurrunt incommoda, quae a duodeno iusto magis ampliato, de quo nobis potissimum sermo est, efficiuntur ac augentur. Quamvis enim aorta ad vertebras lumbares descendens, molem duodeni eiusque pressum valida sua actione superet, vena cava tamen ab incumbente hoc intestino expanso comprimitur. Quod praecipue evenire solet, si voraces post largam lautamque coenam mox ad somnum disponuntur, et supino potissimum, vel in dextrum latus inclinato corporis situ, progressum alimentorum, ex duodeno ad ieiunum difficilem experiuntur. Retardatio quidem sanguinis, in vena cava inferiori facta, superatur tandem, si in somno minus tranquillo situs corporis diversimode mutatur, et hoc quoque modo transitus alimentorum in intestino duodeno adiuvatur, nunquam tamen sine noxa et depravata sanguinis mixtione oriunda, cum eo tempore eveniat, ubi chylus, sanguini admiscendus, libere per cor et pulmones reliquasque corporis partes progredi debet. In systemate venae portae difficilior adhuc est sanguinis ex infimi ventris visceribus morbilibus ad hepar regressus, et uti praeternaturales intestinorum distensiones eundem impedire possunt, sic forte prae reliquis huius canalis partibus ad duodenum nostrum respiciendum erit, quippe, quod nimis expansum venam mesaraicam magnam, sub pancreate egredientem, et supra duodenum ductam, comprimit, § VII idque in loco, qui trunco venae portae vicinus est, et paulo superius magnam venam splenicam suscipit. Quam ob rem iure affirmare possumus, regressum sanguinis a nostro intestino magis, quam ob aliis cohiberi posse. In ipso sanguinis progressu per duodenum non

CHAPTER TWENTY-FOUR

"The Duodenum in its excessive dilatation compresses blood vessels"

It is most certain that blood, as it moves through the abdominal vessels, can be slowed down in many ways, and a variety of diseases, especially chronic ones, will supply such an opportunity. The portal vein system is quite unique and entirely distinct from the vena cava, since it can complete the return of the blood through the abdomen to the heart successfully; however, many problems can occur along the course that it transits, which can be produced and made worse by the greatly widened duodenum, which is our main topic of conversation. Although the descending aorta at the level of the lumbar vertebrae goes over the bulk of the duodenum and compresses it by its strong action, the vena cava, on the other hand, can be compressed by this incumbent and expanded intestine. This occurs especially if the voracious subject, after a large and sumptuous dinner, decides to go to sleep, preferably in a supine position or reclined on the right side of the body, so that the progress of the food from the duodenum to the jejunum experiences difficulty. The delay of blood that occurs especially in the inferior vena cava is overcome, in the end, if during a less tranquil sleep the position of the body is modified in different ways and this also helps the transit of the food through the duodenum, however, never without creating a harmful and corrupted mixture of blood, since it is during this time that the chyle mixing with blood needs to proceed freely through the heart, lungs and other parts of the body. In the portal vein system, the return of blood from the mobile bowels of the lower abdomen back to the liver is more difficult, and thus the unnatural distentions of the intestines may impede this. Perhaps our attention should be more directed to our duodenum than other parts of the intestinal canal because, indeed, when excessively expanded, it compresses the great {superior} mesenteric vein as it exits beneath the pancreas and courses over the duodenum (see chapter 7), and this in a place that is near the

leve commodum accedit, quod duodenalis vena, a trunco venae portae oriunda, immediato nexu cum altera, a mesaraica parva progrediente, coniungatur, eodem modo uti duodenum ipsum excurrat, et unicam quasi venam referat.

trunk of the portal vein, and a little above where it receives the great splenic vein. For this reason, we can affirm that the return of blood can be hindered by the duodenum more than others. In the flow of blood itself through the duodenum one does not derive any slight advantage, because the *duodenal vein*, arising from the trunk of the portal vein, is joined in direct connection with a second one proceeding from the small {superior} mesenteric vein, and runs in the same direction as the duodenum itself, and returns as if it were a single vein.

....

Regarding the **duodenal vein/s**, the situation is a bit more complex than the one described by Claussen. There exist several branches associated with the duodenum:

1) **Posterior superior pancreatico-duodenal vein** that empties directly into the portal vein;

2) **Anterior superior pancreatico-duodenal vein** that empties into the right gastro-epiploic vein, and thus indirectly into the superior mesenteric vein and then into the portal vein;

3) **Anterior inferior pancreatico-duodenal vein** that empties into the superior mesenteric vein and thus indirectly into the portal vein;

4) **Posterior inferior pancreatico-duodenal vein** that empties into the superior mesenteric vein and thus indirectly into the portal vein.

§. XXV.

"Duodenum ampliatus nervos tendit"

Motus sanguinis per abdomen, facile retardandus, inter alias causas, quibus ille adiuvari solet, praecipue nervos habet, quorum magna copia in plexibus sic dictis, a ganglio semilunari sive solari progredientibus, et cum aortae ramis excurrentibus, conspicitur. His enim ad membranosam tubi intestinalis fabricam perductis, motus peristalticus efficitur, et, si opus est, incitatur. Quemadmodum vero ex his motus adminicula cognoscuntur, sic simili ratione a nimia irritatione spasmi et inde excitati dolores explicandi sunt, quippe qui adeo validi redduntur, ut non tantum digestionis negotium turbetur, sed haud raro quoque stases inflammatoriae inducantur. Inter alias intestinorum partes duodenum quoque his incommodis obnoxium est; dum enim illud, nimis expansum, partes vicinas comprimit, nervosi etiam plexus et filamenta inde deducta, quae hac vi comprimi nequeunt, ad minimum ita tenduntur, ut spasmi, et duodeno et partibus connexis molesti, oboriantur. Copiosi inprimis nervi, qui a pari octavo seu vago iuxta oesophagum in ventriculi orificium superius tendunt, per curvaturam parvam quoque ad pylorum usque protrahuntur, ibique praeter nexum universalem paris vagi et intercostalis, filamenta quoque nonnulla a ganglio semilunari dextri lateris, et ipso plexu hepatico suscipiunt, qui prope flexum duodeni primum iuncti tantis tensionibus afficiuntur, quantas saepe prope cardiam observamus. Praeter consensum, quo totum duodenum et plexus renalis dexter his paribus adiacens laeditur, ad ultimum quoque duodeni flexum respiciendum est, quippe, qui novas propagine nervosas a plexu mesenterico superiori, cum renali quoque communicante, recipit, et ideo saepe simul cum prioribus partibus tenditur. Ex quibus omnibus colligitur, alimentorum remoram, in sano duodeni statu summe necessariam, § XX per spasmos obortos, adeo saepe molestam reddi, ut dolores tensivos universales viscerum abdominalium insigniter augeat.

CHAPTER TWENTY-FIVE

"The distended Duodenum stretches nerves"

The flow of blood through the abdomen, which can easily slow down, is mainly helped, among other factors, by nerve fibers, of which a great number are observed in the so-called plexuses, proceeding from the semilunar or solar ganglions and following the branches of the aorta. These reach the muscular layer of the intestinal tube, induce peristaltic motion and, if necessary, excite it. Indeed, just as the benefits from this movement are known, then with a similar reasoning the spasms and the consequent pain from excessive irritation are explained, for these are rendered very strong, so that not only is the process of digestion disturbed but inflammatory conditions are also often introduced. Among the various parts of the intestines, the duodenum is also subject to these disturbances; for when it is expanded too much, it compresses the neighboring organs as well as the nerves and the plexus, and hence the slender filaments, although not compressed by this force, become stretched, so that painful spasms arise both in the duodenum and its adjacent organs. Many nerves arise from the pair of eighth or vagus nerves along the esophagus at the superior opening of the stomach and extend along its lesser curvature to the pylorus. The pair of vagus nerves are connected to both the intercostal nerves as well as to some filaments from the right semilunar ganglion and the hepatic plexus, which are all joined near the first curvature of the duodenum and are affected by extreme tensions, as we observe more often near the gastric cardia. In addition to the consensus that the whole duodenum and the plexus of the right kidney are damaged by these adjacent organs, attention must also be given to the last duodenal flexure which, in fact, receives new nerve branches from the superior mesenteric plexus, and communicates with the renal plexus; therefore, it is often stretched together with the former organs. From all of this we conclude that the troublesome delay of the food, which is greatly necessary in the healthy state of the duodenum (see chapter 20), is often brought about by spasms that arise to such an extent as to cause a remarkable increase in the general tensive pain of the abdominal viscera.

CLAUSSEN'S NEW ILLUSTRATIONS

The two drawings included in Claussen's thesis are reproduced in the following pages. To be exact, these are the ones used in Sandifort's 1788 publication of the dissertation. Figure 1 corresponds to the drawing made with the artist situated at the feet of the cadaver, while figure 2 was created with the artist at the right side of the body.

Claussen emphasized significant differences between the two images, and yet there does not seem to be much variance except for the lesser degree of interference by abdominal organs and peritoneal tissue in figure 2, especially after removal of the right colon. There is consequently an improved highlighting of the duodenum's full course and its relationship with nearby organs.

The true value of these drawings lies in the realistic portrayal of the duodenum and its surrounding organs within the abdominal cavity. Claussen's colleague, George C. Reichel, did a formidable artistic job in depicting these, and his endeavor did not go unnoticed by the medical community of the time.

Figure 1.

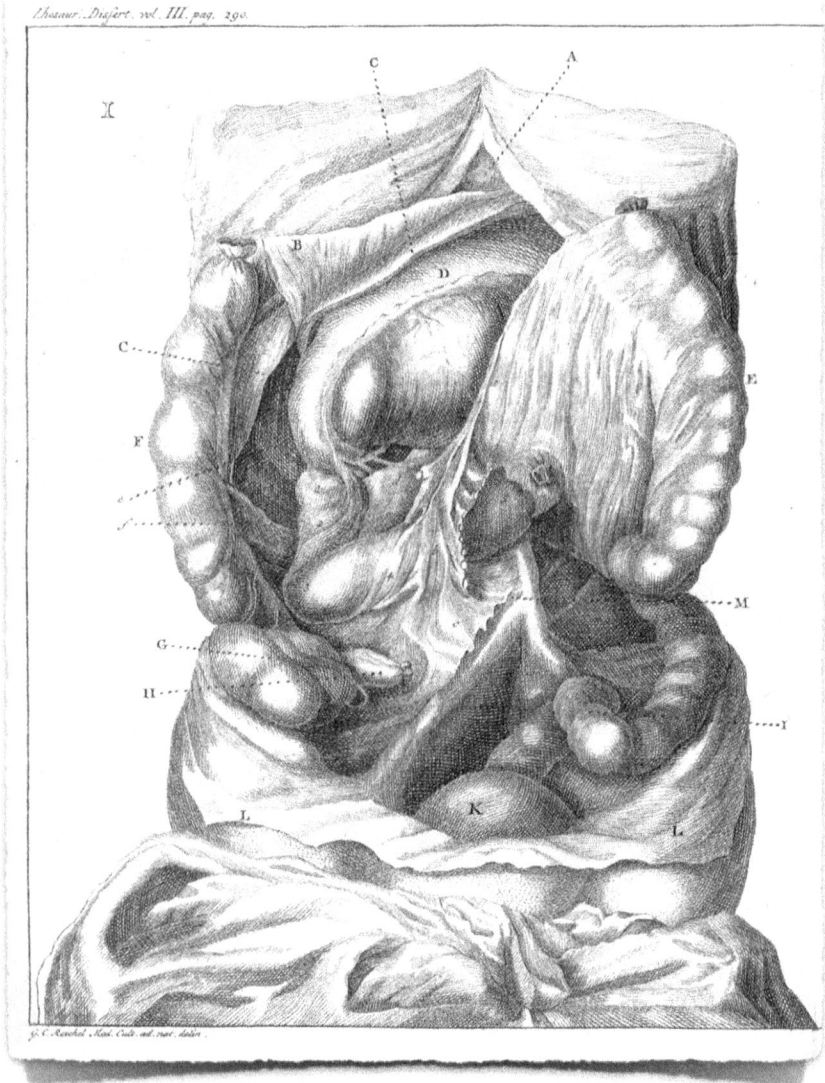

Figure 1 is titled:

"Duodenum cum partibus vicinis in vero situ sistit, delineatore ad pedes constituto."

[Duodenum situated in its true location with nearby organs, with the artist seated at the feet.]

A = xiphoid cartilage;

B = peritoneum;

C = left hepatic lobe;

D = greater curvature of the stomach;

E = left colon;

F = right colon;

G = cecum;

H = terminal ileum;

I = colon towards the rectum;

K = urinary bladder;

L = peritoneum;

M = aorta.

a, b & c = mesentery;

d = head of pancreas;

e = gallbladder.

1 = first curvature of the duodenum;

2 = descending portion of the duodenum;

3 = second or inferior curvature of the duodenum;

4 = ascending and transverse portion of duodenum;

5 = third curvature of the duodenum as it continues into the jejunum.

Figure 2.

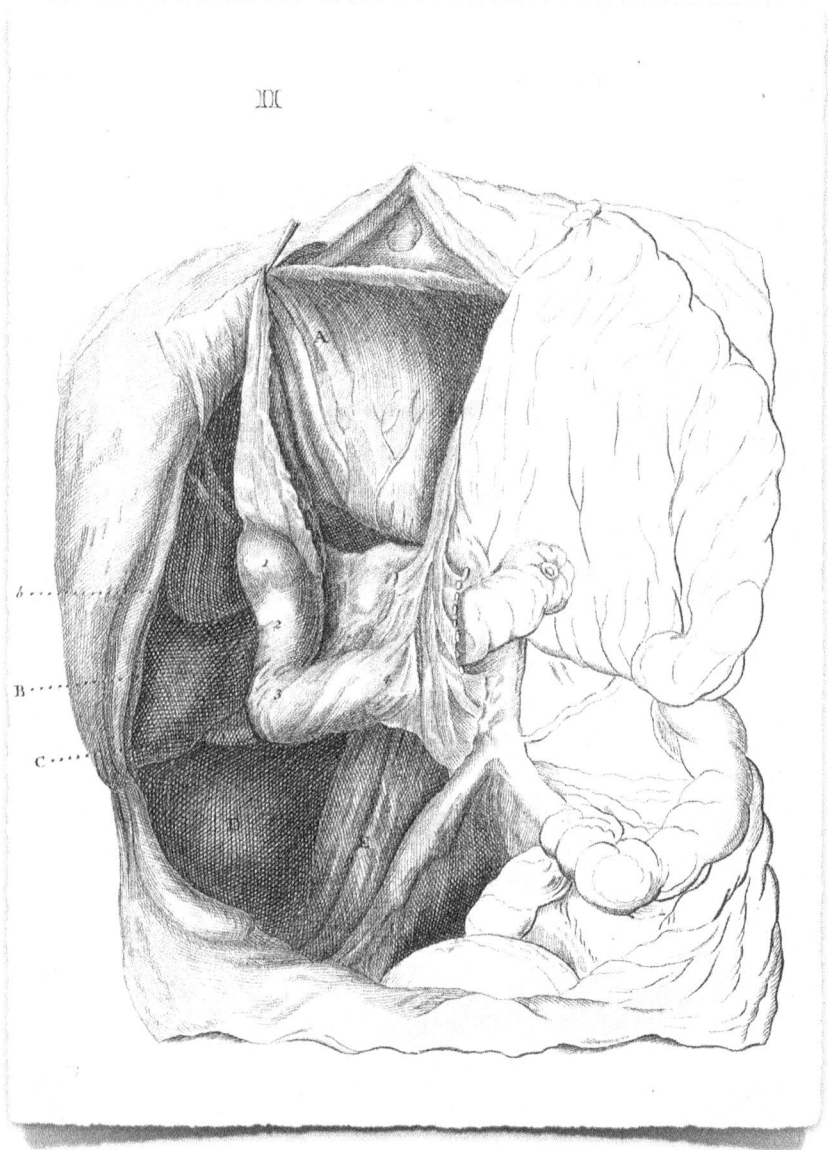

Figure 2 is titled:

"Plurimas partes iconis praecedentis in eodem situ sistit, alias vero a latere dextro visas melius explicat."

[Many organs of the preceding figure at the same location, better explained by another view from the right side.]

A = stomach;

B = right hepatic lobe;

C = lower pole of the kidney;

D = internal iliac muscle;

E = greater psoas muscle.

b = gallbladder.

1 = pyloric fold and duodenal curvature, not obvious in the preceding figure;

2 = descending portion of the duodenum;

3 = lower curvature of the duodenum;

4 = ascending or so-called lower transverse portion of the duodenum.

LEGACY OF CLAUSSEN'S DISSERTATION

After his graduation, it does not appear that Claussen continued with any academic studies pertaining either to the field of the gastroenterology or, for that matter, to the duodenum in particular.

At the time of his graduation, Claussen was the subject of at least two brief introductory monographs which exclaimed honor and praise from his colleagues. One was *Viro Nobilissimo Et Experientissimo Lavrentio Clavssen Hafniensi Dano Doctoris Medici Titvlvm In Illvstri Vniversitate Lipsiensi* written in verse by the Icelandic Paullus **Vidalinus** (the Younger) and published on November 12, 1757. The other was *De Studio Historiae Medicinae iucundo non minus ac utili, ad virum praenobilissimum experientissimum Laurentium Claussen, Havniensem Danum, philosophiae et medicinae baccalaureum, epistola* by Johann Job **Bartsch** and published on October 21, 1757. They both add little to no information about Claussen.

Claussen's work on the duodenal anatomy was, however, referenced by several anatomy authors of the time, first of all by the great Albert von Haller, and enjoyed much recognition throughout the first half of the 19th century. But by the middle of that century his labor was overshadowed by newer works and his dissertation would end up in oblivion.

In 1758, Claussen's thesis was reviewed in *Commentarii de Rebus in Scientia Naturali et Medicina Gestis*. In this series, the anonymous reviewer summarized Claussen's disseration as follows:

"Intestinum duodenum a variis iam auctoribus depictum fuit, nullus vero satis accuratam iconem intestini huius exhibuit. Situs enim illius absconditus difficilem delineationem efficit. Ad pedes cadaveris collocatus delineator, flexum praecipuum sub hepate a latere vesiculae felleae versus renem dextrum deductum nunquam apte cognoscet. Si vero a cadaveris dextro latere positus tractum hunc attendit, tunc finis duodeni versus principia ieiuni excurrens vel obscure, vel in situ obliquo conspicitur. Hisce difficultatibus occurrere tentat Cl.

{Clarissimo} Auctor. Ligaturam nimirum ieiuno ad pollicis distantiam ab eius origine iniecit, ut ventriculus et duodenum flatu distendi potuerint. Reclinatis deinceps quodammodo visceribus vicinis omnem duodeni tractum consideravit, zona coli quoque in media parte divisa. Est vero duodenum prima intestinorum tenuium pars, quae ob flexum suum alimenta semidigesta ad tempus retinet, lentius movet et digerit, celerique postea progressu in ieiuni convolutiones propellit. Flexus duodeni tres sunt, quorum primus et superior a pyloro oritur, cum quo versus corpora vertebrarum reflectitur, sursum tendit et fere cardiae altitudinem attingit. Hoc flexu facto, mox descendit et a latere vesiculae felleae ad renem accedit. Prope superficiem huius interiorem flexus secundus sive inferior positus est et in progressu intestinum coecum attingit, quod quidem in prima icone non perspicitur, in altera vero, ubi partes vicinae remotae sunt, extremitas renis inferior in conspectum venit, et ita descensus duodeni ad coecum comprobatur. Adscendit deinde intestinum nostrum, in mesocoli radice ad pancreatis confinia deducitur, et flexum tertium format, qui in ieiunum continuatus finem duodeni a principio suo non multum distare declarat. Addit tandem Cl. Auctor argumenta quaedam physiologica usum duodeni spectantia et morbos ex illius laesione oriundos erudite considerat."

[The duodenum has already been depicted by various authors, but nobody has displayed an accurate enough illustration of this intestine. Its hidden position, in fact, makes for a difficult drawing. The artist placed at the feet of the cadaver never properly recognizes the main flexure running under the liver and along the side of the gallbladder towards the right kidney. If in fact he directs his attention to this tract when seated on the right side of the cadaver, then the end of the duodenum extending towards the beginning of the jejunum will be observed either indistinctly or in an oblique position. The illustrious Author tries to counter these difficulties. He places a tie on the jejunum at one inch from its origin so that the stomach and the duodenum can be inflated with air. Thereafter, he examines the entire duodenal tract after bending back to a certain extent one after another of the nearby organs and dividing the colon in its middle. The duodenum is indeed the first section of the small intestines, which on account of its flexures detains the semidigested food for a while, slowly moving and digesting it, and then rapidly propelling it into the loops of the *jejunum*. The flexures of the duodenum are three, of which the first and superior one begins at the pylorus, after which it bends towards the vertebral bodies, going upwards and almost reaches the level of the gastric cardia. Having made this bend, it then descends and from the side of the gallbladder reaches the kidney. The second or inferior flexure is situated near the kidney's medial surface and in its course, it reaches the cecum, which indeed is not observed in the first illustration but only in the second, where the neighboring organs have been removed, and the inferior extremity of the kidney comes into view, thus confirming that the *duodenum* descends towards the cecum. The *duodenum* then ascends and enters the root of the mesentery adjacent to the pancreas where it forms the third flexure that continues into the *jejunum*, and this explains how the end of the *duodenum* is not very distant from its beginning. The illustrious Author also adds some physiological arguments pertaining to the function of the duodenum and learnedly reflects upon the diseases arising from its injuries.]

This review is not only a complete summary of Claussen's work but is also written in a much more understandable Latin when compared to the original text of Claussen.

In 1758, Claussen's thesis was again reviewed in the March 13th issue of *Göttingische Anzeigen von Gelehrten Sachen*. His book received a long review which said:

"Eine von Herrn Lorenz Claussen aus Coppenhagen zur Erhaltung der Doctor-Wurde in der Arzneykunst vertheydigte Probtschrift de intestini duodeni situ et nexu, verdient allerdings eine Anzeige, ob sie gleich schon den ein und zwanzigsten October vorigen Jahrs gehalten worden. Es ist bekannt, wie schwer es seye, von der Lage des Zwölffingerdarms eine deutliche Beschreibung, und noch schwerer eine genaue Abzeichnung zu geben, weil er in seinem natürlichen Zusammenhang mit andern Theilen so versteckt liegt, dass man ihn auf keine Weise zu Gesicht bekomt; und doch hingegen gar zu leicht seine natürliche Lage und Gestalt verliehrt, wenn die nahgelegene Theile, die mit demselben verbunden sind, und die hier mit Fett angefüllten Verlängerungen des Darmfells, die es zwar verbergen, aber auch in seiner Lage erhalten, weggenommen werden. Es ist also kein Wunder, wenn man bisshero noch keine recht genaue Zeichnung davon erhalten, und Herr Claussen, der dieser Probschrift zwey Figuren beygefügt, gesteht selbst, es seye fast nicht möglich, eine ganz vollkommene Zeichnung davon zu geben. Nach seinen mit allem Fleiss angelstellten Untersuchungen steigt zwar dieser Darm von dem untern Ende des Magens etwas in die Höhe, doch niemahlen so hoch, dass er der obern Magen-Oefnung gleich kame, beugt sich sogleich wieder herunterwärts, geht unter der Leber an der GallenBlase hin, fast biss an den Blinddarm herunter, und steigt mit einer starken Beugung wieder aufwärts gegen die linke Seite hin, so dass er fast völlig die Höhe seines Anfangs von dem Magen-Pförtner erreicht, da er endlich auf dem Knorpel zwischen dem lezten Rücken- und ersten Lenden-Wirbelbein sich in den leeren Darm endigt. Nach dieser Beschreibung bemerkt er den Ort, wo von aussen die Lage dieses Darms kan bestimmt werden, und zeigt, dass also ein Schmerz, der von der achten Rippe an biss an die Niere sich herunter ziebt, meistentheils in dem Zwölffingerdarm zu vermuthen seye. Da dieses Eingeweyde, ohnerachtet es zwar aus seiner Stelle nicht weichen kan, sich doch, weil es überall mit Fett umgeben ist, sehr stark ausdehnen last, so erklart Herr Claussen, wie durch ein allzustarkes Ausdehnen die nahgelegene Theile und Blutgefasse gedrückt, und die dahin gehende nerven empfindlich gespannt, und heftige Zückungen hierdurch können verursacht werden. Den Nutzen dieses Darms überhaupt und seiner verschiedenen Beugungen insbesondre hat er ausführlich und genau vorgetragen. Die Lage und besonders die untere Beugung dieses Darms hat er durch zwey saubere Zeichnungen zu erläutern gesucht, wobey alle nahgelegene Theile so viel möglich in ihrer natürlichen Lage gelassen worden. Die erstere Zeichnung stellt diesen Darm und die benachbarten Theile so vor, wie sie erscheinen, wenn der Zeichr zu den Füssen des Körpers steht, da in der zweyten Zeichnung die Figur dieser Theile bey der vorigen Lage und Zubereitung abgebildet wird, die fich zeigt, wenn man sie von der rechten Seite her betrachtet.

In dem Anschlag erläutert Herr D. Hebenstreit das acht und zwanzigste Capitel des neunten Buchs der Anecdotorum des Aetii Amideni, welches von dem Ileo und Chordapso handelt, durch die hieher gehörige Stellen aus den Schriften der übrigen alten Aerzte."

[A certain Mr. Lorenz Claussen from Copenhagen defended his thesis *De intestini duodeni situ et nexu* in order to become a Doctor in the Art of Medicine, which certainly deserves an announcement as it was previously held on the 21st of October of last year.

It is well known how hard it is to give a clear description of the position of the duodenum, and even more difficult to furnish an accurate drawing, because whichever way you look at it, it is fairly hidden by its natural relation with other organs. And yet, its natural position and shape are too easily lost if one should remove the nearby organs connected to it as well as the fat-filled extensions of the intestinal walls which not only conceal it but also maintain it in position. So, it is no wonder if one does not obtain a fairly accurate drawing of the duodenum, and Mr. Claussen himself, who included two figures in this thesis, admits that it is almost impossible to render a perfect drawing of it. According to his investigations, which he performed with much diligence, this intestine rises somewhat from the lower end of the stomach, but never so high that it should reach the upper gastric opening, it bends right down again, goes below the liver along the gall-bladder, almost down to the cecum, and rises in a strong bend upwards again towards the left side, so that it almost reaches the level of its beginning by the stomach entrance {cardia}, as it finally terminates in the jejunum at the level of the cartilage between the last thoracic and the first lumbar vertebral bone. According to this description, he noticed the place where the position of this intestine could be determined from the outside, and thus shows that pain which strikes from the eighth rib down to the kidney can be presumed to originate mostly in the duodenum. Even though this bowel cannot be moved from its location, it can become distended because it is surrounded everywhere with fat, so explains Mr. Claussen, and the nearest organs and blood-vessels become stretched and the susceptible sensory nerves undergo tension, thereby causing violent cramps. He has accurately and in detail presented the benefits of this intestine and its various curves. He has endeavored to elucidate the position and especially the lower flexure of this intestine by two accurate drawings, in which all the nearby organs have been left in their natural position as much as possible. The first drawing shows this intestine and the neighboring organs as they appear when the artist is at the feet of the body, while in the second drawing the shape of these organs is depicted during the previous situation and preparation while being viewed from the right side.

At the end, Dr. Hebenstreit explains the 28th chapter of the ninth book of the *Anecdotorum Aetii Amideni* which deals with the *ileus* and *chordapso*, from the corresponding passages of the writings by other ancient physicians.]

In 1758 again, Claussen's thesis was reviewed in the November 10th issue of the *Neue Zeitungen von Gelehrten Sachen auf das Jahr 1757.*

"Herr Lorenz Claussen, Philosof. & Medic. Baccal. aus Copenhagen, vertheidigte am 21 October, zu Erhaltung der Höchsten Würde in der Arzneywissenschaft, seine Probschrift: De intestini duodeni situ & nexu, welche bey Breitkopf auf 4 1/2 Bogen gedrucht, und mit 2 Kupfern gezieret ist. Der geschickte Herr Werfasser, welcher die Zerglieder kunst mit eben demzenigen Fleisse, als die übrigen Theile seiner Kunst, viele Jahre getrieben, und in siener Vaterstadt sowohl, als in Berlin, und bey uns, die besten Lehrmeister zu Vorgängern gehabt hatte, sahe ganz recht ein, dass die Kenntniss von dem Bau derer Theile nicht hinreichend sey, deren Verrichtungen und Nutzen zu bestimmen, wenn man nicht zugleich von ihrer Lage und Verbindung mit andern Theilen auss genaueste unterrichtet ist. Da er demnach den grossen Nutzen erwog, welchen der Zwölffingerdarm zur Verdauung derer Speisen verschaffen muss; und dass gleichwohl die besten Zergliederer, theils diesen Darm nicht genau beschrieben und abgebildet haben, theils aber in Ansehung dessen Lage und Gränzen uneinig sind: so achtete er es der Mühe werth, ass neue diesen Theil in seiner natürlichen Lage und Verbindung sorgsältig zu beschreiben, und abzeichnen zu lassen, welches letzere von seinem guten Freunde, Heren M. Reichel, einem würdigen Candidato Medicinae, mit gehöriger Aufmerksamkeit, in zierlichen Abzeichnungen verrichtet worden ist. Es finden sich zwar sehr viele Schwürigkeiten, wenn man gedachten Darm gehörig untersuchen und abzeichnen will; jedoch giebt und der Herr Verfasser die Art und Weise an, wie solches noch am besten zu bewerkstelligen ist. Er zeiget dahero an, wie das duodenum von dem Ende des Magens abwärts gegen die Niere, und von dar an wieder aufwärts zum Anfange des ieiuni, gehet, ohne gegenwärtig auf die besondern Biegungen mit Fleisse Acht zu haben, welches eher nicht geschehen konnte, als biss er von den Theilen, welche uber, neben, und unter diesem Darme liegen, geredet, und die eigentlichen Gränzen desselben, als worüber eine grosse Uneinigkeit unter denen Zergliederern ist, bestimmet hatte. Die Theile, welche mit erwähnten Darm gränzen, sind der Magen, die Leber, die grosse Schlag- und Blutader, die Adern und das Fett des Gekröses, das Netz, die Gallenblase, die Magendrüse, die Nieren auf der rechten Seite, deren eineige sich mit ihm verbinden, einige nur bloss anliegen. Hierauf wird von den drey besondern Biegungen dieses Darms insbesondere gehandelt, davon die erste gleich unter dem Ende des Magens, die zweyte unter der rechten Niere, und die dritte da befindlich ist, wo das Duodenum beym Ausgange aus dem mesocolo in das ieiunum sich endiget. Man kann aus dieser beschriebenen Lage die Empfindungen dieses Darms von aussen erkennen, und von den Zufällen anderer Theile entscheiden; welches um desto sicherer geschehen kann, da derselbe einen bestandig festen Sitz hat, welcher aber den freyen Durchgang derer Speisen, und die übrigen Verrichtungen desselben, nicht hindert, ob es gleich nöthig ist, dass die Nahrungsmittel durch seine Biegungen einigermassen aufgehalten werden, damit hierdurch sowohl, als durch die einfliessende Galle, und den Magendrüsensaft, die im Munde und Magen angefangene Veränderung und Verdauung der Speisen hierinnen vollendet werde. Gleichwie aber dieser Nutzen des Duodeni vortrefflich, und von sehr grossen Umfange ist; also ist hingegen das Ungemach gleich gross, wenn e entweder von den anliegenden Theilen, die durch Krankheiten selbst verdorben sind, Gewalt leidet, oder selbst über die Gebühr ausgedähnet und sonst schadhaft wird. Der gelehrte Herr Werfasser hat mit eben dem Fleisse die verschiedenen üblen Zufälle deselben erzählt und erklärt, mit welchen er dessen nattürlichen Zustand und Verrichtungen beschrieben hat. Seine Landesleute werden glücklich seyn, welche sich in Zukunft seiner geschickten Vorsorge anvertrauen werden.

Die Einladungsschrift zu dieser feyerlichen Handlung verfertigte Herr D. Joh. Ernst Hebenstreit, Fac. Med. Decanus, als Procancellarius. Sie führet den Tittel: Aetii Amideni ανεκδοτων Lib. IX. caput XXVIII exhibens tenuioris intestini morbum, quem Ileon & Chordapsum dicunt: Una cum veterum super hac aegrotatione sententiis, auf 2 1/2 Bogen. Unser in den Alterthümern und denen alten griechischen Schriftstellern vorzüglich gelehrte Herr Werfasser, hatte von denen in griechischer Sprache noch nicht herausgekommenen Büchern des Aetius jüngsthin eine Probe abdrucken latzen, welche von denen in dieser Art Gelehrten sehr wohl aufgenommen worden ist. Vortetzo erwählet er aus eben diesem Schriftsteller wiederum ein Stück, welches noch nicht gedruckt ist, und mit der Abhandlung des Herrn Candidati eine Aehnlichkeit hat. Es begreift dasselbe die Krankheit in denen kleinen Gedärmen ausser dem Zwolffingerdarm, welche die Alten Ileon und Chordapsum nennen. Man findet dieses Uebel, mit seinen verschiedenen Arten, Ursachen, und Mitteln darwider, in dieser alten Schrift umständlich, sorgfältig, und natürlich abgehandelt. Es ist dem griechischen Texte eine lateinische Uebersetzung, zum Gebrauch dererienigen beygefügt, welche die Urschrift ohne solche nicht überall verstehen möchten, die aber genau nach dem griechischen eingerichtet ist. In dem angehängten Commentario werden die schwehren Stellen erlä, und ähnliche Stellen aus andern Griechen angezogen. Wir münschen, mit auswärtigen Gelehrten, unserm unermüdeten Herrn Verfasser, Gefundheit und Kräfte, damit die von ihm versprochene, und so lange gewünschte Ausgabe der übrigen Bücher des Aetius, durch seinen Fleiss bald erscheinen möge."

[Mr. Lorenz Claussen from Copenhagen, Baccalaureate in Medicine and Philosophy, defended on the 21st of October, towards the achievement of the highest honors in Medicine, his thesis *De intestini duodeni situ and nexu*, which is printed on four wide half-sheets and engraved with two copperplates. The skilled author who for many years has performed the art of dissection with just the same diligence as the rest of his skills, both in this city and Berlin, and like us has had the best teachers, was well aware that the knowledge of the structure of these organs was not sufficient to determine their functions and usefulness, unless at the same time their position and connection with other organs were precisely determined. Therefore, since the duodenum furnishes great assistance to the digestion of food and due to the fact that the best prosectors on the one hand have not accurately described and depicted this intestine and on the other are in disagreement with regard to its position and limits, he considered it worthwhile to newly describe the natural position and connection of this organ and to draw it, the latter of which has been executed with elegant illustrations and with due attention by his good friend, Mr. Reichel, a worthy candidate in Medicine. There are, of course, very many difficulties when one wishes to properly examine and draw this intestine; but the author of the book shows us the way in which this can still be accomplished. He therefore demonstrates how the duodenum goes downwards from the end of the stomach to the kidney, and from there goes up again to the beginning of the jejunum, without at this time having to pay attention to the particular bends, which would not happen until he had determined the actual limits of those organs which lie above, beside, and under this intestine, for which there is great disagreement among the Anatomists. The organs which surround this intestine are the stomach, the liver, the aorta and vena cava, the veins and the fat of the mesentery, the omentum, the gallbladder, the pancreas, and the right kidney, some of which are attached to it, and some only abut it. This is followed by the three unique bends of this intestine, of which the first is situated just below the end of the stomach, the second under the right kidney, and the third where the duodenum terminates at the end of the mesocolon. From this described position, the symptoms of this intestine can be recognized externally, and the consequences on other organs may be determined; this may be more evident since it has a firm position, which does not prevent the free passage of the food nor the rest of its functions, and does not hinder the need for the passage of food to be slowed down by its bends to a certain degree so that the transformation and digestion of the food, that began in the mouth and stomach, are accomplished in this way also by the inflowing bile. Just as the usefulness of the duodenum is admirable and of a high degree; on the other hand, it may be equally damaged when it is violated by the adjoining organs which are involved by diseases themselves, or if it is itself excessively stretched and otherwise damaged. The erudite gentleman author, with the same diligence with which he has described the natural condition and functions of the latter, has narrated and explained the various nepharious accidents. His countrymen will be happy, and in the future will entrust themselves to his skillful intuition.

The introduction to this solemn event was made by Dr. Johann Ernst Hebenstreit, Dean and Vice-Chancellor of the Faculty of Medicine. It has the title: *Aetii Amideni Anekdoton Lib. IX. caput XXVIII. exhibens tenuioris intestini morbum, quem Ileon & Chordapsum dicunt: Una cum veterum super hac aegrotatione sententiis*, on 2 half-sheets. Our author {Hebenstreit}, who is particularly learned in the antiquities and ancient Greek writers, had recently been able to print a sample of the books of Aetius, which had not yet been published in the Greek language, which was very well received by scholars of this topic. He chose a piece from the same author, which has not yet been printed, and has a resemblance to the treatise of the candidate. It deals with the disease in those small intestines, besides the duodenum, which the Ancients called Ileon and Chordapsum. This illness, with its various types, causes, and means, is carefully and naturally dealt with in this ancient treatise. The Greek text is translated into Latin, without which the original text cannot be understood everywhere, and is arranged exactly according to the Greek. In the attached commentary, the passages are explained, and similar passages are drawn from other Greeks. We, along with foreign scholars, wish our indefatigable writer health and powers that the promised, and so long desired, edition of other books of Aetius may soon appear through his diligence.]

In 1764, Albert von **Haller** wrote the 7th volume of his *Elementa Physiologiae Corporis Humani*. He referenced Claussen several times when discussing the anatomy of the duodenum, such as in the following phrases:

"duodeni . . . quae est intestini tenuis pars a pyloro ad mesocolon transversum producta"
[the duodenum . . . which is the part of the small intestine that proceeds from the pylorus to the transverse mesocolon];

"tunc flexu facto emergit antrorsum, & insigniter sursum, & per foramen proprium, in quo mesocolon transversum & nascens mesenterium conveniunt descendit in partem abdominis; quae est infra mesocolon transversum."
[having then made the curvature it emerges forward, and notably upward, and through its own hole, in which the transverse mesentery and the origin of the mesentery come together, it descends into the abdomen; which is below the transverse mesocolon.]

"Rectum nullo modo dici potest,"
[In no way can it be said to be straight]

"Inde pergit, primumque ductum suum perficit in universum transversum ut dextrorsum tendat, & modice retrorsum; tamen ut una alternis imperfectis flexionibus, antrorsum emergere & retrorsum se recipere, adsendere sursum, & deorsum redire videatur; quo ad vesiculae fellis cervicem adtingat."

[From there it continues and completes its first channel in an entire transverse direction so that it tends to the right and slightly backward; however, as a second imperfect flexure, it appears to emerge forward and retreat backward, to ascend upwards and return downwards to where it touches the gallbladder neck.]

"In pleno ventriculo potius paulum descendit."
[In a full stomach it descends a little more.]

"Transversum ergo sinistra repetit, pone pancreas, pone venae portarum principem truncum, & pone arteriam mesentericam"
[Transverse and then returns to the left, behind the pancreas, behind the main trunk of the portal vein, and behind the mesenteric artery.]

In 1776, Albert von **Haller** wrote the *Bibliotheca Anatomica qua scripta ad Anatomen et Physiologiam Facientia a Rerum Initiis Recensentur*. When citing Claussen's dissertation he briefly said:

"Bona disp. {disputatio} cum icone propria. Ab origine paulum adscendere, inde descendere in coecum usque intestinum: tunc iterum sinistrorsum pene ad originem suam ex pyloro redire: in cartilagine sub ultima vertebra dorsi finiri."
[A good discussion with its own drawings. From the beginning it ascends a little, to then descend as far as the cecum, then again to the left to return almost to its original source from the pylorus: to end at the cartilage under the last vertebra of the back {i.e., T-12}.]

The phrase "cum icone propria" is both a compliment to Claussen and also a little dig to those anatomy books which included illustrations that were mere copies of anatomical drawings made by others, especially those by Vesalius.

In 1778, the 3rd edition of the *Encyclopédie ou dictionnaire raisonné des sciences, des arts et des métiers* added new references to the history of anatomy which included Claussen:

"Laurent Claussen a donné une bonne these sur le duodenum."
[Laurent Claussen has offered a good thesis on the duodenum.]

In 1778, Eduard **Sandifort** wrote the 3rd volume of his *Thesaurus Dissertationum, Programmatum, Aliorumque Opusculorum Selectissimorum ad Omnem Medicinae Ambitum Pertinentium* in which he reprinted Claussen's entire dissertation with its figures.

158

Sandifort's presentation to Claussen's dissertation is reprinted on page 5.

In 1780, Eduard **Sandifort** wrote the text-atlas *Tabulae Intestini Duodeni*. On page 15, he purports to reprint chapter #5 from Claussen's dissertation, but actually ends up paraphrasing parts of chapters #5 and #6:

"Duodenum adscendit, reflectitur, & descendit: primo hoc tractu retrorsum magis quam sursum vergere videtur, porro extrorsum potius ad vesiculam felleam usque progreditur, ejusque cervicem & partem corporis tegit, a qua sub hepate recta quasi via defertur, & ad interiorem renis dextri marginem, seu hylum renalem, deducitur, in pinguedine delitescens; ab hac parte renis introrsum ac sursum vergit, & ita oblique super spinam dorsi vel potius super venam cavam & arteriam aortam in sinistram cadaveris partem deducitur. Hoc in tractu pinguis mesenterii radix supra duodenum extenditur, illudque quasi abscondit, ut, nisi extensione quadam facta, observari nequeat, a parte tamen sinistra huius radicis mesenterii duodenum magis in conspectum prodit, quam a dextra, cum in progressu suo dextrum hoc latus relinquat. Dem autem hoc suo tractu ad ultimam dorsi vertebram ascendat, ad mesocolon quoque pertingit, & in duplicatura ejus a radice sua pingui suscipitur, a qua tamen, levi flexu facto, tandem in inferiore parte secedit, & antrorsum in jejunum continuatur. Ultima haec duodeni curvatura sub pancreate haeret, haec enim glandula extremitate sua sinistra & cuspidata a liene aliqua ex parte deorsum extenditur."

[The duodenum ascends, turns around, and descends. This first segment appears to turn backwards more than upwards, and then continues outwards mainly towards the gallbladder, of which it covers the neck and part of its body. From here it then proceeds downwards under the liver in an almost straight line, and extends towards the inner margin, or *hilum*, of the right kidney while being concealed in adipose tissue. Thereafter, the duodenum proceeds partly inward and upward from the aforementioned kidney, and then extends obliquely across the dorsal spine, or rather over the vena cava and the aorta. The duodenum then passes to the left side of the body. Along this route, the fatty root of the mesentery courses over the duodenum and almost hides it, such that, had a certain distention not been made, it would not be observed. However, as it leaves the right side, the duodenum comes more into view on the left side of the mesenteric root, more so than on the right. But while it ascends up to the last thoracic vertebra, it also reaches the mesocolon, and is received into its fold by its fatty root, from which, after making a slight curve, it finally exits in the lower part, and continues forward into the jejunum. This last curvature of the duodenum hugs the bottom of the pancreas, whose pointed left extremity extends almost to the dorsal aspect of the spleen.]

In 1791, Friedrich **Hildebrandt** wrote volume 3 of the *Lehrbuch der Anatomie des Menschen*. When discussing the anatomy of the duodenum, he cited only two books, Claussen's and Sandifort's. His comment about Claussen's book was:

"Eine gute Beschreibung, der zwo gute Abbildungen beigefugt sind."
[A good description; two good illustrations are enclosed.]

In 1793, Justus Christian **Loder** wrote *Anfangsgründe der medicinischen Anthropologie und der Stats-Arzneykunde*. Upon discussing the anatomy of the small bowel, including the duodenum, he recommended consulting five authors, one of which was Claussen (Claussen – 1757; Sandifort – 1780; Santorini - posthumous; Helvetius – 1721; Albini – 1722 & 1724).

In 1795, Johann Friedrich **Blumenbach** authored the 2nd edition of *Anfangsgründe der Physiologie*. This would be later translated into English under the titles of either *The Elements of Physiology* or *The Institutions of Physiology* (1820). On page 226, Blumenbach mentions the *Zwölffingerdarm* (literally, the "twelve-finger intestine") and in a footnote makes reference to Claussen's thesis, its reprint by Sandifort, as well as Sandifort's own book *Tabulae Intestini Duodeni*.

In 1809, Friedrich Ludwig **Augustin** wrote the *Lehrbuch der Physiologie des Menschen: mit vorzüglicher Rücksicht auf neuere Naturphilosophie und comparative Physiologie*. He included Claussen's book as one of three references regarding the anatomy of the duodenum (Albini – 1724; Ludwig – 1754; Claussen – 1757).

In 1815, Jean-Nocolas **Marjolin** wrote volume 2 of the *Manuel d'Anatomie*. Upon discussing the anatomy of the duodenum, he recommended consulting five authors, one of which was Claussen (Brunner – 1715; Santorini - 1724; Clausen (*sic*) – 1757; Sandifort – 1780; Bleuland - 1789).

In 1816, John and Charles **Bell** wrote the 4th edition of *The Anatomy and Physiology of the Human Body*. In a footnote for the chapter on the anatomy of the duodenum, the authors said:

"See a good description of the duodenum by M. Laurent Bonazzoli, in the Transactions of the Academy of Bologna. And the Dissert. L. Claussen. de duodeni situ et nexu. Sandifort Thes. V. III. Monro, Medical Essays."

In 1817, G. D. **Yeats** read at the College of Physicians of London "Some Observations on the Duodenum, with Plates descriptive of its situation and connections" which was published in 1820 in *Medical Transactions*. Considering the title of his article, much of Yeats' discussion not surprisingly revolved around the words of Claussen, and even one of the two plates included was copied directly from Claussen's own drawing. Yeats said of Claussen's book:

"A detached essay or two have in former periods been published; one valuable one by Hoffman, replete with many excellent remarks, and another by Dr. Claussen in 1757."

Yeats also added that:

"Bonazzoli indeed and Hoffman and Monro and Claussen have given us some very valuable observations, both practical and anatomical, but the path they have laid open has been untrodden."

In other words, the teachings of Claussen and others have not had the deserved impact, and other organs such as the noble liver have maintained a predominant role as the culprit of diseases which, as Yeats stresses, "have most certainly taken their origin from a morbid condition of the duodenum itself."

In 1820, Johann Friedrich **Meckel** wrote volume 4 of his *Handbuch der Menschlichen Anatomie*. This book was translated into French (1825), Italian (1826) and English (1832). In the chapter on the anatomy of the duodenum, he included Claussen's book as one of two bibliographical references (Claussen – 1757; Sandifort – 1780).

In 1825, Carlo Francesco **Bellingeri** wrote the monograph *Storia delle Encefalitidi che furono epidemiche in Torino nell'anno 1824* [History of the Encephalitides that were Epidemic in Turin in the year 1824]. In the addendum entitled "Storia di duodenite acuta susseguita da encefalitide" [A case of acute duodenitis followed by encephalitis] Bellingeri ended the discussion by saying:

"Nelle infiammazioni interne se manca il dolore della parte infiammata, manca quel segno, che con certezza guida il Medico al riconoscimento della flogosi, e della sua sede. Altronde per riconoscere la duodenitide dalla sede del dolore si esige molta avvertenza; e su ciò si potrà consultare quanto scrisse Claussen nella sua disertazione: De intestini duodeni situ et nexu inserita nel Thesaurus dissertationum di Sandifort, t. 3, p. 283."

[Among the internal inflammatory processes, if pain is lacking in the inflamed organ, then the signs that guide the physician to the recognition of the inflammatory process and its site of origin are missing. Moreover, a lot of precaution is necessary in order to identify duodenitis as the site of the pain's origin, and in this regard one may consult the writings of Claussen in his thesis *De Intestini Duodeni Situ et Nexu* included in Sandifort's *Thesaurus Dissertationum*, volume 3, page 283.]

In 1826, Annibale **Omodei** compiled volume 38 of the *Annali Universali di Medicina*. He included a review of the previously mentioned book by Bellingeri from 1825 and, in order to substantiate this reference, Omodei included a transcription of half of chapter 15 "De duodeni situ externe definiendo disseritur" from Claussen's book.

In 1826, John **Bostock** wrote volume 2 of *An Elementary System of Physiology*. In a footnote he said:

"The duodenum has been named by some anatomists the ventriculus succenturiatus, or accessory stomach, as being the organ in which the process of chylification appears to be perfected. Claussen, de duodeno, in Sand. Thes. t. iii. p. 273."

In 1827, Martin **Münz** wrote volume 3 of the *Handbuch der Anatomie des menschlichen Körpers*. Claussen was cited as one of six important published references regarding the anatomy of the duodenum.

In 1833, volume 9 of the *Encyclopädisches Wörterbuch der medicinischen Wissenschaften* was published. Claussen was cited as one of nine references regarding the anatomy of the intestines, alongside such greats as Albini, Brunner and Peyer.

In 1836, Ernest-Alexander **Lauth** wrote the *Handboek der Practische Ontleedkunde*, translated into French as *Nouveau Manuel de l'Anatomiste*. In the chapter about the "Twaalfvingerigen Darm" [duodenum], he included Claussen's book as one of two relevant bibliographical references (Claussen – 1757; Sandifort – 1780).

In 1841, Joseph **Frank** (1771-1842) wrote volume 3 of the 2nd edition of his *Praxeos Medicae Universae Praecepta*. This was translated into Italian in 1844 as "Patologia Interna: Malattie del Tubo Intestinale" in the *Enciclopedia delle Scienze Mediche*. When discussing the diseases of the intestinal canal, he briefly introduces its anatomy and says:

"in sex partes dividi solet, quarum primam sistit duodenum [4]), quod parvus ventriculus dictum"

[it is usually divided into six parts, the first of which is the duodenum, also called the small stomach]

The [4]) footnote refers to Claussen's dissertation from 1757 and its reprint in Sandifort's book from 1780.

In 1841, volume 17-18 of the *Répertoire générale des sciences médicales au XIXe siècle: Dictionnaire de médecine et chirurgie pratiques* was published. In the entry for the "**Intestin**," it included Claussen's dissertation as one of five references (Vater – 1720; Monro – 1738; Claussen – 1757; Sandifort – 1780; Bleuland - 1789) when dealing with the anatomy of the duodenum.

And, finally, in 1968 Claussen's book was briefly cited in the *Berner Beiträge zur Geschichte der Medezin und der Naturwissenschaften*. This stated:

"Lorenz Claussen [publ. 1757] veröffentlichte eine gute Abhandlung über das Duodenum."

[Lorenz Claussen published in 1757 a good treatise on the duodenum.]

. . . .

And so, for almost an entire century Claussen's treatise on the topographical anatomy of the duodenum was held in such high esteem as to be one of the few recommended texts where medical students could deepen their knowledge of this hidden organ. Because of his in-depth study, Claussen's name was associated with some of the great anatomical luminaries of the time. His book, however, was never translated into another language. At the time, Latin was the international language through which European scientists communicated with each other. Once Latin was no longer in favor, Claussen's book became scientifically obsolete and was replaced by new texts, and thus almost completely forgotten.

FINAL REMARKS

Claussen's dissertation tackles several aspects of the topographical anatomy of the duodenum. One aspect has to do with the true configuration and course of the duodenum. Another addresses its relationship with certain nearby organ structures, most importantly the mesocolon and the root of the mesentery. Part of the issue stems from the complexity of the juncture between these three abdominal structures. Claussen would have resolved it more readily if he had presented the embryological development of this site. In fact, the mesentery, mesocolon and mesoduodenum all begin as one membranous structure that evolves as follows: (1) the mesoduodenum disappears as the duodenum becomes adherent to the retroperitoneum; (2) the mesentery extends to allow for the elongation of the remaining small bowel; and (3) the mesocolon becomes reshaped by the rotation of the ileocecal junction from the upper left to the lower right abdomen. In this process, the mesocolon twists above and around the mesenteric root and crosses over the duodenum.

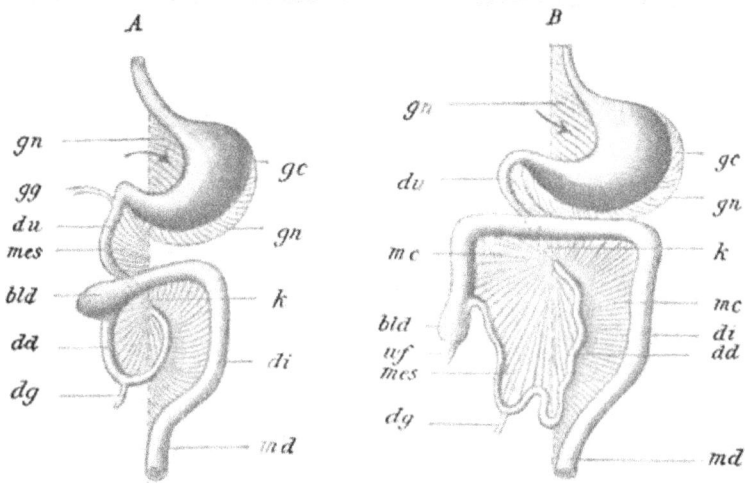

Diagram of the embryological development of the mesenteries. du = duodenum; mes = mesentery, including mesoduodenum; mc = mesocolon; bld = cecum.

The result is the concealment of the duodenum, not only behind the peritoneum of the posterior wall of the abdominal cavity but also by the heavily adipose tissue of the root of the transverse mesocolon and by the superior end of the mesenteric root. The right-most end of the gastrocolic ligament is sometimes seen adhering to the proximal portion of the duodenum.

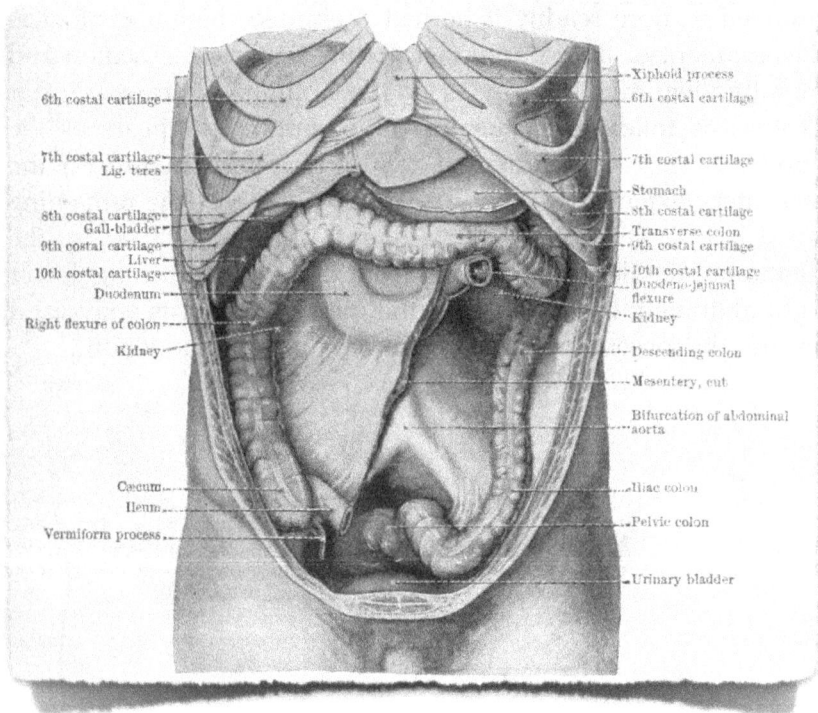

The paths of the root of the transverse mesocolon and that of the mesentery are depicted within the following drawing (see *boxed* captions) wherein the concealing organs have been removed.

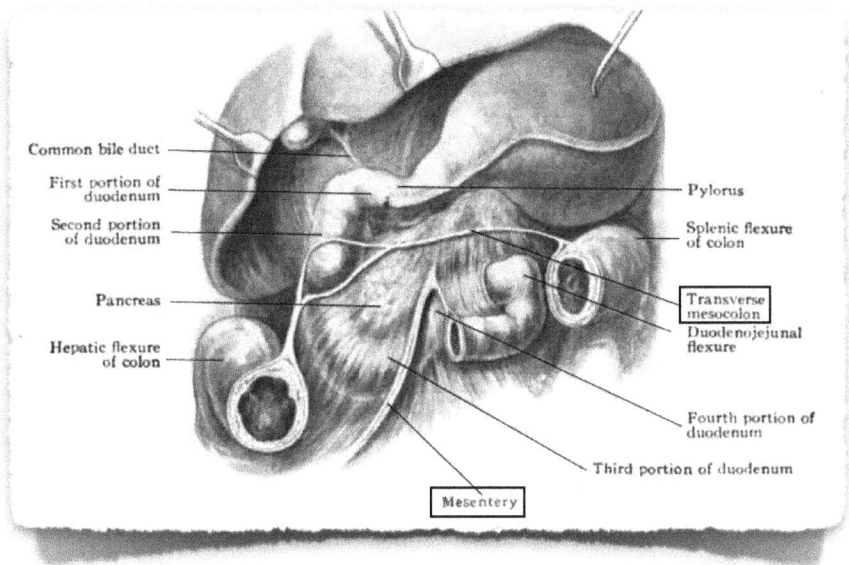

Common bile duct

First portion of duodenum

Second portion of duodenum

Pancreas

Hepatic flexure of colon

Pylorus

Splenic flexure of colon

Transverse mesocolon

Duodenojejunal flexure

Fourth portion of duodenum

Third portion of duodenum

Mesentery

As for the title of the dissertation, it had some predecessors such as in Winslow himself who used a similar title: "Situation, Connexion" for the second paragraph in his description of the duodenum in his book *Exposition Anatomique de la Structure du Corps Humain* (1732) (see page 38).

Also, much of the text written by Claussen had been previously summarized in the 1738 paper by Monro "The Description and Uses of the Intestinum Duodenum." Indeed, a comparison of these two works makes one realize how important Monro's work was and how Claussen was able to expand on it with his thesis and at the same time incorporate most of the published knowledge of the duodenum. It appears that a significant impetus for Claussen was, in fact, the lack of a proper and realistic drawing that could accompany any description of this hidden segment of small bowel.

In summary, Claussen's monograph is significant for several reasons:

(1) It introduces us to the scholarly world of medical education of mid 18th century Europe, and demonstrates his own experience within the fluid system of teaching curricula among various universities;

(2) It summarizes the story of the duodenum and the long road this digestive organ had to travel before gaining a proper anatomic description;

(3) And finally, it presents us with the first lifelike image of the duodenum drawn by an experienced artist who paid special attention to its true form and relation with neighboring organs.

APPENDIX

The following is a transcription of the Lorenz Claussen's life presented by Professor Hebenstreit at the end of his monograph *Aetii Amideni Anekdoton liber IX* and written at the same time as Claussen's dissertation of 1757.

. . . .

"Atrocis istius morbi non minimam sensionem duodenum quoque habet; Illud enim, postquam calamitas ad summum profecta est, merdam, cui, ad sedem abiturae via dempta nunc fuit, in se recipit et ventriculo, euomendam, tradit, et miserabili officio, pro vita aegroti utcunque tuenda, ac morbo trahendo, cum sanari ille plerumque haud possit, fungitur. Gratulor Clarissimo Candidato meo de erudita huius intestini descriptione, qua effecit, ut omnes intelligant id, quod nos privatis experimentis de eodem captis satis superque intelleximus, eum honoribus academicis proxime secuturis esse dignissimum.

Quem cum nunc sum laudaturus, eumque, tanquam praeclarum exemplum vitae academicae recte ac laudabiliter actae, ad imitationem propositurus, opus est, quo res ab illo bene gestas altius repetam, et, quanta animi contentione, ab ineunte aetate, omnes illas eruditionis partes, quae medicum faciunt ornantque, emetiri ac penitus ediscere enisus fuerit, enumerem: Natus est Iuvenis Clarissimus Dominus LAURENTIUS CLAUSEN, Hafniensis Danus, Philosophiae et Medicinae Baccalaureus, in Metropoli Regni Daniae, Hafnia, scientiarum ac liberalium artium ac Summorum Virorum, qui artem salutarem, et superiori et hoc, quo nunc vivimus, seculo, experimentis ac literarum monumentis auxerunt, et adhuc dum augent, felici patria. Nascenti illi, omnes, quae vitam beatam polliceri ac facere poterant, commoditates Parentum conditio ac fortuna tradidit: Pater enim illi est, Vir Ampliss. Heningus Ditlef Clausen, Aulae Regiae Chirurgus et artis obstetriciae Magister, in utraque arte, celebris. Mater vero, Matrona, omnibus, quae sexum ornant, virtutibus conspicua, AGNETA, ex gente nobili SCHIOETT.

Etiamsi autem isthac felicitate, quae sane insignis est, cum possumus ab illis, qui nos genuerunt, educari, usus diu haud fuit, siquidem Utroque Parentum, mature defunctis, sub teneriore aetate sua orbatus est, nihil tamen deesse illi opportunitatis passa divina providentia fuit, quem Avus maternus, BALTHASAR PAULSEN SCHIOETT, et, illo ex vivis cedente, Avia materna ELISA MARIA, VIDVA SCHIOETT, magno amore comprehensum, ita coluerunt, ut, se Parentibus orbum esse, haud sentiret, et Honestissima isthaec Matrona, ex nepote suo, reduce, non leve senectutis, quam Deus tranquillam, durabilem, faxit, solatium est habitura. Veluti filii, plerumque, ad vitae genus eligendum, exempla a patribus capere solent, ita quoque ad istam Patris sui imaginem educans NOSTER statim fuit atque illi, qui Parentum loco tunc erant, teneram eius aetatem ita formaverunt, quo et Christianae Religionis cognitio ex quisita illi traderetur, et literarum, quarum capaces sub primo vitae stadio sunt animi, primae ducerentur lineae. Igitur sollertissima cura factum est, ut, tum talis educatio, qualis civem, patriae olim utilem futurum, docet, tum praeceptores probi doctique illi contingerent, quorum privata institutione ita literarunt elementis imbutus fuit, ut inter cives urbis patriae academicos Anno MDCCXLVIII, recipi posset, quo facto, ex more et institutis illius Academiae, privatum praeceptorem sibi eligit Virum Illustrem IO. PETRUM ANCHERSEN, Iuris utriusque Doctorem, Potentissimo Danorum Regi a Consiliis Iustitiae, et Eloquentiae Professorem, cuius etiam, in humanioribus literis, lectiones, deinde, in Logica et Metaphysica, THESTRUPII, in geometricis scientiis, ZIGENBALGI, in Astronomia, HORREBOVII, iunioris, in Physica experimentali, amicissimi HEII, Virorum in sua quorumlibet disciplina celeberrimorum, praecepta sedulo audivit, illisque, pro praestita sibi opera, publicas, me interprete, decernit gratias, atque ea felicitate in istis philosophiae partibus progressus NOSTER est, ut, post examen, uti in celebri isthac Academia, mores maiorum hoc secum afferunt, solemne, debitos illi honores haberet AMPLISSIMUS PHILOSOPHORUM ORDO et publico diligentiae ac eruditionis scholasticae testimonio ornandum censeret esse discipulum eiusque profectus Laudabiles publice depraedicaret, quod praeconium, ex Universitatis Hafniensis instituto, optimum est, et Candidatis bene merentibus, laudis ergo, tribuitur. Superato isthoc tentamine, primam lauream philosophicam a Praeceptoribus in se collatam, eiusdem anni mense Iunio, tanquam rerum academicarum porro omni. Studio agendarum incitamentum, habuit. Hinc suadentibus amicis, proprio una impetu delatus, ad medicinam et chirurgiam discendam animum adplicuit et quidem in anatomia et chirurgia lectiones hibernas illustris CRYGERI audivit, ex cuius familiaribus colloquiis ac institutione privata, in variis novis fasciarum chirurgicarum generibus recte applicandis, proficere, ipsiusque benevolentia singulari frui, Nostri licuit: Simili etiam favore, experientissimi VOHLERTI, chirurgi aulici, cuius in se merita nunquam satis praedicari a se posse, fatetur, nec non Amicissimi HENINGSONII, quorum laboribus assiduum spectatorem sociumque praestitit, atque ab ipsis, quoties rariora artis exempla occurrerent, admissus liberaliter fuit, usus est. Experientiss. HEVERMANNUM in theatro medico anatomiam explicantem, audivit, atque a Celebri Viro CAPPELIO metallurgiam, Experientiss. LODBERGO FRISIO et EICHELIO, materiam medicam, BUCHWALDO et OEDERO botanicam, accepit atque huic, in itinere per Saelandiam scaniamque, historiae naturalis excolendae causa suscepto, comitem sese adiunxit. Hinc anno MDCCLIII, Berolinum profectus, Praeceptorem Fautoremque insignem nactus est Virum in arte salutari peritissimum MECKELIUM, quo duce per duo semestria hiberna anatomiam et physiologiam, ad ductum Illustris HALLERI, didicit, artemque obstetriciam illi debet. Experientiss. GLEDITSCHII opera, in ediscenda materia medica inque privato botanicae studio, institutus, praeterea e lectionibus et operationibus chirurgicis, in Nosodochio Regio, quod a Cartitate dicitur, PALLASII, POTTII, MUZELII et laudati GLEDITSCHII, clarissimorum in sua quorumlibet arte Virorum, profecit. Postea Anno MDCCLV., mutata studiorum sede, ciuibus illustris Philureae nostrae, Magnifico WINCKLERO Rectore, adscriptus, Fautores et Praeceptores novos nactus est, quibus docentibus varias medicinae partes ulterius atque penitius cognoscere illi datuit. Doctrinam de electricitate, ipso WINCKLERO tradente, itemque summorum Virorum, CHRISTII et GELLERTI lectiones, super elegantioribus litteris, audivit. Excellentissimor. Virorum, BOSII in ediscenda Physiologia, inque herbis exoticis contemplandis, in historia artis medicae, PLAZII, in chemiae et pharmacologiae studio QUELLMALZII et HUNDERTMARKII, in exercitiis disputatoriis, HEINII, institutionibus usus, artem salutarem ad unguem edidicit, quem perpolivit Hospes et Magister benevolus Excellentissimus LUDWIGIUS, illi pathologiam et therapiam ac rem litterariam medicam explicans, quem et ego inter amicos et discipulos habui, eumque, in praxi medica inque chirurgia, anthropologia forensi, mineralogia et zoologia, perlustratis eam ob causam Musaeis Richteriano et Linckiano, comite studiorum FRATRE eius dulcissimo, BALTHASARE CHRISTIANO, Iuris utriusque Candidato doctissimo, performavi, atque illum domesticis exercitii gratia institutis tentaminibus, excellentis ingenii Iuvenem comperi, siquidem, me quoque Praeside, dissertationem de medicamentis, ut menstruum agentibus, ad leges chemicas, contra aliter sentientium obiectiones, palam defendit, cui si quid a me benefactum est, gaudeo. Hisce igitur omnibus ex voto peractis, die XXX. Iulii huius anni, tentamen artis medicae, quod theoreticum dicunt, sustinuit et RECTE respondit, proximo abhinc mense per totam hebdomada, pro Licentia summos in arte medica honores impetrandi, de aquarum, quae foetum ambiunt, theoria et praxi, lectiones, audientibus Commilitionibus, ac Praeceptoribus, habuit publicas. Mox, die X. Sept. alterum acriusque tentamen praxeos medicae cum laude sustinuit et RECTE respondit, die instante XXI. Octobris, pro summis in arte medica honoribus capessendis de intestini duodeni situ et nexu, sine Praeside, quod illi in honorem cedit, et Mandato Regio iussum est, sub Moderamine Viri Magnifici Excellentissimi D. ANTONII GUILIELMI PLAZII, Physiologiae Professoris Publici, Academiae Naturae Curiosorum Collegae, disputaturus, et publicum vitae academicae recte ac laudabiliter gestae testimonium, applaudentibus omnibus bonis, consecuturus. Etiamsi autem bonae merces praecone haud indigent, iustum tamen est ac aequum, publice laudari Iuvenem, qui, tanto studio, per tot experimenta in diversis Academiis edita, ad summam laudem contendit. Illud a me factum iri, meque, Candidati mei nomine, summo Numini pro concessis illi ad res bene agendas viribus, hinc Praeceptoribus, Fautoribus ac Amicis, pro studio ac labore in bene merentem collocato, publicas acturum gratias, ipsi autem Candidato, me, habita oratione, de hypothesium valore ad veritatem investigandam, iusta persoluturum esse, significo, et, quo huic panegyri interesse velint, Rectorem Magnificum, Ilustrissimos Comites, Utriusque Reipublicae Proceres, Generosissimos atque Nobilissimos Commilitones, decenter invitatos cupio. Dedi Lipsiae, ex Facultate Nostra, Dominica XIX. post Festum Trinitatis, die XVI. Octobris Anni a reparata nobis per Christum Dominum Nostrum salute MDCCLVII."

This *curriculum vitae et studiorum* of Lorenz Claussen contains a wealth of names of his family, his teachers and the luminaries of the medical society of the time. His academic exposure was quite impressive and included professors of the universities of Copenhagen, Berlin and Leipzig.

Hebenstreit begins by introducing the medical student Lorenz Claussen who was born in Copenhagen within the Kingdom of Denmark. His father was the surgeon and obstetrician Hening Ditlev Claussen and his mother Agneta Schiøtt. The young Lorenz was to become orphaned at an early age and would thereafter be raised by his benevolent maternal grandparents Balthasar and Elisa Schiøtt.

In 1748, Lorenz Claussen entered the University of Copenhagen where he received teachings from professors Anchersen (literature), Thestrup (logic and metaphysics), Ziegenbalg (geometry), Horrebow (astronomy) and Hee (experimental physics). Having past his exams, he decided to study medicine and began by attending the anatomy and surgery lessons of professor Krueger. His studies continued with the teachings of professors Wohlert (surgery), Hennings (surgery), Heuermann (anatomy), Cappel (metallurgy), Friis (medicine), Eichel (medicine), Buchwald (botany), and Oeder (botany). Claussen accompanied the latter two professors on a journey to Zeeland and Skåne (in southern Sweden) in order to improve his knowledge of natural history.

Then, in the year 1753, he set out for the University of Berlin, where he attended classes of anatomy and physiology given by the notable teacher Meckel, and obstetrics by the illustrious Haller. Other teachers were Gleditsch (medicine, surgery, botany), as well as Pallas, Pott and Muzel who taught the art of surgery at the Charité Hospital of Berlin.

Thereafter, in the year 1755, he changed his seat of studies to the University of Leipzig. Claussen attended classes by a variety of noteworthy professors such as Winkler (electricity), Christ (literature), Geller (literature), Bose (physiology, botany, history of medicine), Plaz (chemistry, pharmacology), Quellmalz (research), Hundertmark (research), Hein (healing), and Ludwig (therapy, pathology, medical literature, medical practice, surgery, forensic anthropology, mineralogy, and natural history). In order to perfect his knowledge of these subjects Claussen also visited the Richter and Linck Museums of Leipzig.

It was in Leipzig that he met up with his younger brother Balthasar Christian, a student of law.

On December 11, 1756, Claussen presented the paper *De medicamentis, ut menstruum agentibus, ad leges chimicas*. Then, on July 30, 1757, he sustained and passed with flying colors a major theoretical exam on the art of medicine. A month later he held lessons on *De aquarum, quae foetum ambiunt, theoria et praxi* [On the theory and practice of fluids that surround the fetus]. Soon after this, on September 10, he took and passed the practical exam of medicine.

And thus, on October 21, 1757, Lorenz Claussen presented his thesis *De intestini duodeni nexu et situ*. In place of the dean, the dissertation was moderated by the Anton Wilhelm Plaz, professor in General Physiology and member of the Academy of Sciences Leopoldina.

The following is a brief description of all the individuals mentioned by Hebenstreit.

- - - - - - - - - - - - - -

Johannes (Hans) Peder **Anchersen** (1700-1765) was a Danish historian and philologist. In 1736 he became a professor at the University of Copenhagen and thereafter Dean and Rector where he gave numerous dissertations on classical philology.

Christian **Thestrup** (1689-1750) was a professor of Logic and Metaphysics at the faculty of Philosophy in the University of Copenhagen.

Ernst Gottlieb **Ziegenbalg** (1716-1758) studied Theology in Jena and, after studying Mathematics for two years in England, became Professor of Mathematics in Copenhagen in 1747.

Peder [Nielsen] **Horrebow** (Horrebov) (1679–1764) was a Danish astronomer and became professor of mathematics at the University of Copenhagen in 1714. He also became director of the university's observatory. Horrebow invented a way to determine a place's latitude from the stars. It is now called the Horrebow-Talcott Method. The crater Horrebow on the Moon is named after him. His son, also an astronomer, was Christian Pedersen **Horrebow** (1718–1776) who succeeded him as director of the observatory associated with the University of Copenhagen. Neith, a supposed moon of Venus, was spotted by Christian Horrebow, while he was studying the passage of this planet from 1766 to 1768. He also discovered the periodicity of sunspots.

Severin **Hee** (1706-1756) was a professor of Medicine at the University of Copenhagen. Christian **Hee** (1712-1781) was a professor of Mathematics and Experimental Philosophy at the Marine Institution / University of Copenhagen.

Simon **Krüger** (1687-1760): see page 14.

Hans Friedrich **Wohlert** (1703-1779) was a Danish surgeon who, along with Krüger, laid the groundwork for the art of Surgery in the Denmark of the first half of the 18th century. Born in Kiel, Wohlert came to Copenhagen where he was trained by Henning Ditlev Claussen and took a degree in surgery in 1739. He thereafter worked as a surgeon until 1768. As such, he operated on Queen Louise of Great Britain for a sandwiched hernia in 1751, but he was not able to save her life.

Wilhelm **Hennings** (1716-1794) was a Danish surgeon who studied at the Theatrum Anatomico-chirurgicum in Copenhagen, where he graduated in 1748. At Simon Krügers death in 1760, he became general director of surgery in Denmark (and Norway).

Georg **Heuermann** (1723-1768) was a Danish surgeon who issued the invaluable four-volume work *Physiologie*. He became professor at the University of Copenhagen.

Joachim Diderich **Cappel** (1707-1784) was a German pharmacist, chemist and metallurgist who moved to Copenhagen in 1747 and worked at the Friderich's Hospital.

Christian Lodberg **Friis** (1699-1773) graduated in medicine from Leiden in 1725 and then practiced in Copenhagen where he became professor of medicine in 1747.

Johan **Eichel** (1729-1817) had studied medicine in various German universities and practiced as a country doctor in Odense. Upon his death, he bequeathed a scholarship for students of the University of Copenhagen.

Balthasar Johannes von **Buchwald** (1697-1763) was a surgeon who became professor of medicine in Copenhagen (see also page 1).

George Christian **Oeder** (1728-1791) was a German-Danish botanist, medical doctor, economist and social reformer. He studied medicine at the University of Göttingen under Albrecht von Haller. He then settled as a Doctor of Medicine in the city of Schleswig. The king called him to Copenhagen in 1751 on von Haller's recommendation. The University of Copenhagen, reluctant as it was to employ foreign experts, resisted Oeder's appointment as a Professor Ordinarius. Thus, he was appointed

Royal Professor of Botany and soon led the installation of a new botanical garden. From 1753 he led the publication of a monumental botanical plate work, *Flora Danica*.

Johann Friedrich **Meckel** (1714-1774) was a professor of anatomy, botany, and obstetrics in Berlin.

Albrecht von **Haller** (1708-1777) was a professor of medicine, anatomy, botany and surgery at the University of Göttingen.

Simon **Pallas** (1694-1770) was a German physician and professor of surgery at the Collegium medico-chirurgicum and first surgeon at the Charité Hospital of Berlin.

Johann Heinrich **Pott** (1692-1777) was a German physician and chemist. He was professor of practical chemistry (pharmacy) at the Collegium medicum-chirurgicum in Berlin.

Friedrich Hermann Ludwig **Muzell** (1715-1784)) was a German physician and a pupil of Boerhaave. In 1744 he became professor at the Charité in Berlin.

Johann Gottlieb **Gleditsch** (1714-1786) was a German physician and botanist who studied medicine at the University of Leipzig, and eventually relocated to Berlin as a professor of botany at the Collegium Medico-Chirurgicum.

Johann Heinrich **Winkler** (1703--1770) was a German scholar of wide interests, but best remembered as a physicist. He taught in the same school in Leipzig where Johann Sebastian Bach worked, and even wrote the libretto for Bach's cantata "Froher Tag, verlangte Stunden." He taught at Leipzig University, first as professor of Greek and Latin from 1742 to 1750, and then as professor of Physics from 1750. He was one of the founders of experimental physics at the University of Leipzig.

Johann Friedrich **Christ** (1700-1756) was a German archaeologist and art historian. In 1739 he was awarded a professorship for poetry and the chair of physics at the University of Leipzig.

Christian Fürchtegott **Gellert** (1715-1769) was a German poet and a professor of Philosophy at the University of Leipzig. He was a pupil of J. F. Christ.

Ernst Gottlob **Bose** (1723-1788), a distant relative of Johann Sebastian Bach, was a German botanist and physician. He obtained the professorship of botany at the Leipzig Medical Faculty in 1755.

Anton Wilhelm **Plaz** (1708-1784) was a German physician and botanist. In 1749 he became a full professor of botany at Leipzig, where afterwards he served as professor of physiology (1754-1758), anatomy and surgery (1758), pathology (1759-1773), and therapy (1773-1784).

Samuel Theodor **Quellmalz** (1696-1758) was a German chemist and a professor of anatomy and surgery (1726-1737, 1747-1748), physiology (1737-1747), pathology (1747-1758) and therapy (1757-1758) at the University of Leipzig.

Karl Friedrick **Hundertmarck** (1715-1762) was a German physician. He was appointed professor of medicine in 1748, and associate professor of physiology and of anatomy and surgery in 1754 at the University of Leipzig.

Johann Abraham **Heine** (1723-1792) was a physician at the St. Georgen Krankenhaus, the second largest hospital in Leipzig.

Christian Gottlieb **Ludwig** (1709-1773) was a German physician and botanist. He became a professor of medicine (1747), pathology (1755) and therapy (1758) at the University of Leipzig.

The Richter Museum was a Cabinet of Curiosities in Leipzig established by Johann Christoph **Richter** (1689-1751), a banker and merchant. The cabinet was composed of more than 2,000 specimens of minerals, precious stones, animals, fossils, insects and plants. Johann Ernst Hebenstreit wrote a catalog of the collection in 1743.

The Linck Museum was a Cabinet of Natural History and Curiosities in Leipzig established by Heinrich **Linck** (1638-1717), a German pharmacist and collector.

Balthasar Christian **Claussen** (1731-1789) was the younger brother of Laurentius Claussen. He became a lawyer after studying at the University of Göttingen, which he left in 1755, and finishing his studies in Leipzig. Thereafter he journeyed to Strasbourg, Geneva and Paris. He later worked as a brewer in Copenhagen. Most interestingly, he carried with him a "stambog" or "friendship book," a sort of guestbook popular in the 16th to 18th centuries, in which friends would write quotes, verses, and drawings. In Balthasar's book we meet a number of his fellow students, including Friederich Christian von Haven and Per Forsskål who later lost their lives during their exploration of "Arabia Felix," today's Yemen. This adventure was recently recounted in a 1962 book by the title *Arabia Felix: The Danish Expedition of 1761-1767* by Thorkild Hansen. Balthasar married Hedevig Cathrine Nissen, born Luchtrant or Lochteran (b. 1744) and the couple had three children: Agnethe Cecilia (1768-1835), Christiana Maria (1769-1803), and Hedevig Laurentze (1772-1802). Balthasar's wife Hedevig was previously married to Lorentz Nissen (1713-1765) who died while she was pregnant with their daughter Kristine Lorentzdotter (b. 1765). The couple had two previous children, Elisabeth Sophie Nissen (b. 1744) and Gregoria Hoeg Nissen (b. 1751).

ILLUSTRATION CREDITS

Page 2: Laurentius Claussen, *De Medicamentis ut Menstruum Agentibus ad Leges Chymicas* (1756)

Page 5, 142, 144: Eduard Sandifort, *Thesaurus Dissertationum Programmatum, volume III* (1778)

Page 7: Ernestus Hebenstreit, *Aetii Amideni Anekdoton liber IX, caput XXVII, exhibens tenuioris intestini morbum quem Ileon et Chordapsum dicunt* (1757)

Page 9: Laurentius C. Claussen, *De Tremore* (1754)

Page 11: Laurentius Claussen, *De Intestini Duodeni Situ et Nexu* (1757)

Page 21-22: Io. Gorraei, *Definitionum medicarum libri 24* (1564)
[https://archive.org/details/bub_gb_64G4Aa7RlPQC]

Page 25: Ioannis Riolani Filii, *Opera Anatomicae* (1649)
[https://archive.org/details/bub_gb_TF9tWd706e0C]

Page 26: Ioannis Riolani, *Encheiridium anatomicum et pathologicum* (1649)
[https://commons.wikimedia.org]

Page 28: Frederik Ruysch, *Adversariorum Anatomico-Medico-Chirurgicorum* (1736)
[https://babel.hathitrust.org]

Page 29: "The Anatomy Lesson of Dr. Frederick Ruysch" by Jan van Neck (1683): [https://commons.wikimedia.org]

Page 32: Boerhaave and Albini, *Opera Omnia Anatomica & Chirurgica* (1725)
[https://openlibrary.org]

Page 33-34: Andrea Vesalius, *De humani corporis fabrica libri septem* (1555)
[https://commons.wikimedia.org]

Page 36: Realdo Colombo, *De re anatomica* (1559)
[https://commons.wikimedia.org/wiki/File:De_Re_Anatomica.jpg]

Page 37: Reinier Graaf, *Tractatus Anatomico-Medicus De Succi Pancreatici* (1664)
http://www.biusante.parisdescartes.fr/histmed/image?03552]

Page 39: Jaques-Benigne Winslow, *Exposition Anatomique de la Structure du Corps Humain* (1732) [https://archive.org/details/b21471332_0003]

Page 42-43: René Croissant de Garengeot, *Splanchnologie ou L'Anatomie des Visceres* (1728) [https://archive.org/details/splanchnologieou02gare]

Page 48: Alexander Monro, "The Description and Uses of the Intestinum Duodenum" in *Medical Essays and Observations* (1738).

Page 52: Albrecht Haller, *Iconum Anatomicarum Partium Corporis Humani* (1745), fascicle 2

Page 75: Albrecht Haller, *Iconum Anatomicarum Partium Corporis Humani* (1743), fascicle 1

Page 83: Francis Glisson, *Anatomia Hepatis* (1665)
[https://archive.org/details/bub_gb_9lJkAAAAcAAJ]

Page 93: Thomas Willis, *Pharmaceutice Rationalis sive Diatriba De Medicamentorum Operationibus* (1674)

Page 95: *Cunningham's Text-book of Anatomy* (1914)
[https://archive.org/stream/cunninghamstextb00cunn#page/1150/mode/2up]

Page 105, top: *Cunningham's Text-book of Anatomy* (1914)
[https://archive.org/stream/cunninghamstextb00cunn#page/1158/mode/2up]

180

Page 105, bottom: *Cunningham's Text-book of Anatomy* (1914)
    [https://archive.org/stream/cunninghamstextb00cunn#page/1170/mode/2up]
Page 108-109: Christian Gottlieb Ludwig, *Quo Observata in Sectione Cadaveris Feminae Cuius Ossa Emollita Erant* (1757)
Page 125: Jean-Baptiste de Sénac, *L'Anatomie d'Heister* (1735)
Page 131: Johann Vesling, *Syntagma Anatomicum* (1641)
Page 165: *Morris's Human Anatomy*, part 4 (1907)
    [https://archive.org/stream/morrisshumanana01jackgoog#page/n32/mode/2up]
Page 166: *Cunningham's Text-book of Anatomy* (1914)
    [https://archive.org/stream/cunninghamstextb00cunn#page/1156/mode/2up]
Page 167: Gwilym G. Davis, *Applied Anatomy* (1916), 4th edition (modified)
    [https://archive.org/stream/appliedanatomy00davirich#page/n1/mode/2up]

REFERENCES

1543 - Andrea Vesalius: *De Fabrica Corporis Humani.*

1559 - Realdo Colombo: *De Re Anatomica.*

1564 - Jean de Gorris: *Definitionum Medicarum, Libri XXIIII.*

1618 - Jean Riolan: *Anthropographiae.*

1641 - Johann Vesling: *Syntagma Anatomicum*

1649 - Jean Riolan: *Opera Anatomicae.*

1664 - Reinier de Graaf: *De Succi Pancreatici.*

1665 - Francis Glisson: *Anatomia Hepatis*

1674 - Thomas Willis: *Pharmaceutice Rationalis sive Diatriba De Medicamentorum Operationibus in Humano Corpore.*

1682 - Thomas Willis: *Opera Omnia.*

1688 - Brunner: *De Glandulis in Intestino Duodeno hominis detectis.*

1708 - Boerhaave: *Institutiones Medicae*

1715 - Jacques-Benigne Winslow: "Nouvelles Observations Anatomiques sur la Situation et la Conformation de Plusieurs Visceres" in *Mémoires de l'Academe Royale*

1717 - Lorenz Heister: *Compendium Anatomicum*

1721 - Lorenz Heister: *A Compendium of Anatomy*

1724 - Giovanni Domenico Santorini: *Observationes Anatomicae.*

1725 - Boerhaave and Albini: *Opera Omnia Anatomica & Chirurgica.*

1728 - René Croissant de Garengeot: *Splanchnologie ou L'Anatomie des Visceres.*

1732 - Jacques-Benigne Winslow: *Exposition Anatomique de la Structure du Corps Humain*

1734 - Jacques-Benigne Winslow: *An Anatomical Exposition of the Structure of the Human Body.*

1735 - Jean-Baptiste de Sénac: *L'Anatomie d'Heister, avec des essais de physique sur l'usage des parties du corps humain, & sur le méchanisme de leurs mouvemens*, 2nd edition

1736 - Frederik Ruysch: *Adversariorum Anatomico-Medico-Chirurgicorum*

1738 - Alexander Monro (the *first*): "The Description and Uses of the Intestinum Duodenum" in *Medical Essays and Observations*, volume 4, 2nd edition.

1740 - Albrecht Haller: *Boerhaave's Praelectiones Academicae*, volume I.

1743 - Albrecht Haller: *Iconum Anatomicarum Partium Corporis Humani*, fascicle 1

1745 - Albrecht Haller: *Iconum Anatomicarum Partium Corporis Humani*, fascicle 2

1747 - Jacques-Benigne Winslow: *Esposizione Anatomica della Struttura del Corpo Umano del Winslow.*

1753 - Jacques-Benigne Winslow: *Expositio Anatomica Structurae Corporis Humani.*

1756 - Laurentius Claussen: *De Medicamentis ut Menstruum Agentibus ad Leges Chymicas.*

1757 - Anon: *Neue Zeitungen von Gelehrten Sachen auf das Jahr 1757.*

1757 - Laurentius Claussen: *De Intestini Duodeni Situ et Nexu.*

1757 - Ernst Hebenstreit: *Aetii Amideni Anekdoton liber IX, caput XXVII, exhibens tenuioris intestini morbum quem Ileon et Chordapsum dicunt.*

1757 - Christian Gottlieb Ludwig: *Quo Observata in Sectione Cadaveris Feminae Cuius Ossa Emollita Erant.*

1758 - Anon: *Commentarii de Rebus in Scientia Naturali et Medicina Gestis,* volume 7, part 1.

1758 - Anon: in *Göttingische Anzeigen von Gelehrten Sachen,* March 13.

1758 - Anon: in *Neue Zeitungen von Gelehrten Sachen auf das Jahr 1757*

1758 - Haller: *Disputationes ad Morborum Historiam et Curationem Facientes,* volume 6.

1764 - Albert von Haller: *Elementa Physiologiae Corporis Humani,* volume 7.

1772 - Lorenz Heister: *Compendio Anatomico di Lorenzo Heister.*

1776 - Albert von Haller: *Bibliotheca Anatomica qua scripta ad Anatomen et Physiologiam Facientia a Rerum Initiis Recensentur.*

1778 - Eduard Sandifort: *Thesaurus Dissertationum Programmatum, volume III.*

1780 - Eduard Sandifort: *Tabulae Intestini Duodeni.*

1791 - Friedrich Hildebrandt: *Lehrbuch der Anatomie des Menschen.*

1793 - Justus Christian Loder: *Anfangsgründe der medicinischen Anthropologie und der Stats-Arzneykunde.*

1795 - Johann Friedrich Blumenbach: *Anfangsgründe der Physiologie,* 2nd edition.

1809 - Friedrich Ludwig Augustin: *Lehrbuch der Physiologie des Menschen: mit vorzüglicher Rücksicht auf neuere Naturphilosophie und comparative Physiologie*

1815 - Jean-Nocolas Marjolin: *Manuel d'Anatomie,* volume 2.

1820 - Blumenbach: *The Institutions of Physiology*

1820 - Johann Friedrich Meckel: *Handbuch der Menschlichen Anatomie* - vol 4.

1825 - Carlo Francesco Bellingeri: *Storia delle Encefalitidi che furono epidemiche in Torino nell'anno 1824.*

1826 - Annibale Omodei: *Annali Universali di Medicina,* volume 38.

1827 - Martin Münz: *Handbuch der Anatomie des menschlichen Körpers.*

1833 - *Encyclopädisches Wörterbuch der medicinischen Wissenschaften,* volume 9.

1836 - Ernest-Alexander Lauth: *Handboek der Practische Ontleedkunde.*

1841 - Joseph Frank: *Praxeos Medicae Universae Praecepta,* 2nd edition.

1841 - *Répertoire générale des sciences médicales au XIXe siècle: Dictionnaire de médecine et chirurgie pratiques,* volume 17-18.

1844 - Joseph Frank: "Patologia Interna: Malattie del Tubo Intestinale" in the *Enciclopedia delle Scienze Mediche.*

1887 - Albert H. Buck (ed): *A Reference Handbook of the Medical Sciences,* volume 5

1916 - Gwilyn G. Davis: *Applied Anatomy,* 4th edition

1968 - *Berner Beiträge zur Geschichte der Medizin und der Naturwissenschaften.*

INDEX

184

www.ingramcontent.com/pod-product-compliance
Lightning Source LLC
Chambersburg PA
CBHW060548210326
41519CB00014B/3396